T0396659

Smart Agriculture

Volume 6

The book series Smart Agriculture presents progress of smart agricultural technologies, which includes, but not limited to, specialty crop harvest robotics, UAV technologies for row crops, innovative IoT applications in plant factories, and big data for optimizing production process. It includes both theoretical study and practical applications, with emphasis on systematic studies. AI technologies in agricultural productions will be emphasized, consisting of innovative algorithms and new application domains. Additionally, new crops are emerging, such as hemp in U.S., and covered as well. This book series would cover regions worldwide, such as U.S., Canada, China, Japan, Korea, and Brazil.

The book series Smart Agriculture aims to provide an academic platform for interdisciplinary researchers to provide their state-of-the-art technologies related to smart agriculture. Researchers of different academic backgrounds are encouraged to contribute to the book, such as agriculture engineers, breeders, horticulturist, agronomist, and plant pathologists. The series would target a very broad audience – all having a professional related to agriculture production. It also could be used as textbooks for graduate students.

Zhao Zhang · Xufeng Wang
Editors

Towards Unmanned Apple Orchard Production Cycle

Recent New Technologies

Editors
Zhao Zhang
Key Laboratory of Smart Agriculture System Integration Research, Ministry of Education
China Agricultural University
Beijing, China

Xufeng Wang
Tarim University
Aral, Xinjiang, China

ISSN 2731-3476 ISSN 2731-3484 (electronic)
Smart Agriculture
ISBN 978-981-99-6123-8 ISBN 978-981-99-6124-5 (eBook)
https://doi.org/10.1007/978-981-99-6124-5

This Springer imprint is published by the registered company Springer Nature Singapore Pte Ltd.
The registered company address is: 152 Beach Road, #21-01/04 Gateway East, Singapore 189721, Singapore

Series Editor's Preface

Smart agriculture technology has significantly transformed the world, which does not only increase agriculture efficiency but also make labor intensive agricultural operations automatic. Over the past decades, a huge number of autonomous machinery (or robotics) has been developed and tested to replace human labor for agricultural operations.

Apple is one of the most favourite fruits in the world, and China ranks the first in apple production. During the past decades, however, apple production manner has not been changed at all, and it is almost fully by labor. The top four apple production operations include harvest, in-field sorting, pollination, and bagging. For the harvest, workers need to wear a bucket and take advantage of ladders to access high apples; for the in-field sorting, workers have to manually grade and sort individual apples; for the pollination, workers utilize a brush to manually complete the pollination process; for the bagging, workers wear a bag for each apple. The four operations are now the bottleneck for apple production, since there lacks enough agriculture employee to meet the requirements.

Researchers during the past decades have investigated great efforts to developing robotics to automatize the apple production full cycle, with the four operations (i.e., harvest, in-field sorting, pollination, and bagging) focused. A lot of robotics have been developed for harvesting, in-field sorting, pollination, and bagging. However, there lacks a book to systematically and comprehensively discuss the technology progress, which gap has been filled by this book.

By specifically focusing on automation of apple production cycle, Drs. Zhao Zhang and Xufeng Wang edited this book titled "Towards Unmanned Apple Orchard Production Cycle—Recent New Technologies". This book provides fundamental knowledge, as well as practical application examples for autonomous technology in apple production. This book provides a comprehensive and timely information source for readers who are interested in learning about the important subject area of apple production mechanization.

Zhao Zhang
Professor of College of Information
and Electrical Engineering
China Agricultural University
Beijing, China

Preface

Apple is one of the most popular fruits in the world. Though new technologies have transformed the world significantly with a lot of manual work has been partially or even fully replaced by robots or automated systems, apple production cycle is still heavily labor involved, and the current production pattern has not been changed during the past decades. Shrinking labor pool and increasing labor cost are putting strains on the apple production. Thus, it requires to develop innovative sensing and automation technology for apple production cycle.

Researchers have investigated a lot of efforts on developing automatic systems for apple production cycle during the past decades. Though a lot of academic research outcomes have been published, there lacks a book systematically and comprehensively presenting the research progress, which gap has been filled by this book.

The most labor-intensive operations in apple product include, but not limited to, harvest, bagging, infield sorting, pollination, and pruning. Thus, researchers have put a lot of efforts to develop corresponding technologies. The first chapter reviews the overall automatic technology progress in apple production cycle, followed by the second chapter focusing on apple bagging robots. The third chapter shows new technology on apple infield sorting technology to replace human labors, and the fourth chapter also focus on apple bagging technology, with more efforts on describing the exiting automatic bagging mechanism on the market. Then, the fifth chapter addresses the flower pollination approach using robotics and the sixth chapter discusses the pruning technology progress. Both seventh and eighth chapters describe the apple harvest robotics technology progress, especially the machine vision applied to detect and localize apples. The last chapter comes to the apple flower pollination issue again, and more efforts on describing UAV technology on flower pollination application. With the nine chapters, a systematic technology review of automatic technology on apple production have been introduced.

This book does not only present academic staff state-of-the-art autonomous technology on apple production cycle, but also present layman a general idea on this research area. Editors wish this book can spark readers' interest in autonomous systems for apple production cycle, and promote the technology progress.

Beijing, China
Alar, China

Zhao Zhang
Professor of College of Information and Electrical Engineering

Xufeng Wang
College of Mechanical and Electrical Engineering

Contents

Chapter 1
Developments of the Automated Equipment of Apple in the Orchard: A Comprehensive Review

Mustafa Mhamed, Muhammad Hilal Kabir, and Zhao Zhang

Abstract Agriculture products are essential to the globe since they meet all of the requirements for human sustenance; thus, constantly developing new methods and equipment to increase output and stability is crucial. The apple is one of the most significant fruits grown worldwide. It offers nutritional and health advantages; therefore, it is essential to increase output and ensure quality by creating intelligent tools and equipment. This study analyzes the growth of intelligent, automated apple fruit equipment in five stages: picking, pruning, thinning, pollinating, and bagging. First, summarizing robots, applications, resources, and findings; next, identifying noteworthy advancements and tactics; then, highlighting the significant difficulties; and lastly, outlining potential prospects and our outlook for the future. They all contribute to providing services and maintaining the growth of research communities for increased apple fruit productivity and quality.

Keywords Artificial Intelligence (AI) · Apple fruits · Harvesting · Pruning · Thinning · Pollinating · Bagging

---

M. Mhamed · M. H. Kabir (✉) · Z. Zhang
Key Laboratory of Smart Agriculture System Integration, Ministry of Education, Beijing, China
e-mail: mkhilal@zju.edu.cn

Key Lab of Agricultural Information Acquisition Technology, Ministry of Agriculture and Rural Affairs, China Agricultural University, Beijing, China

College of Information and Electrical Engineering, China Agricultural University, Beijing, China

Z. Zhang
e-mail: zhaozhangcau@cau.edu.cn

Z. Zhang and X. Wang (eds.), *Towards Unmanned Apple Orchard Production Cycle*, Smart Agriculture 6, https://doi.org/10.1007/978-981-99-6124-5_1

1.1 Introduction

Several industries have effectively used robots [1], applications for handling, moving, processing, inspecting, and regulating the quality of materials. In recent years, the concept of mechanization, the use of automated machinery and robotics in agriculture, has gained increased visibility, and there are several examples of successful mechanized farming [2–8]. Robots in agriculture are being used to increase production and food quality while lowering labor and time expenses [9, 10]. The lack of trained human labor in agriculture is one of the primary drivers of robotic farming. It impacts the development of emerging nations [11]; therefore, if farmers are given the tools they need to succeed with the help of robots, the country's agricultural output might significantly increase [12, 13].

Nowadays, harvesting is the major agricultural task that supporting robots are doing. Modern agriculture must devise fresh strategies to boost productivity. One strategy is to decrease and target energy inputs more effectively than before, using in-formation technologies now accessible as smarter machines [14]. With the development of autonomous production systems, we now have the chance to create a whole new line of agricultural machinery based on clever small machines that can act appropriately in the appropriate situation, at the appropriate time, and in the appropriate manner. In developed nations, environmental issues are increasingly playing a bigger role in politics, particularly concerning agriculture [15].

A successful shift to the age of robots will require resolving several challenges with field operations in agriculture, which are quite complicated. A thorough system analysis of the field operation and a cost-benefit analysis are required before developing an automated solution [16, 17]. For such a system to be successful, it must meet several extremely precise characteristics, including low weight, tiny size, autonomy, intelligence, communication, safety, and adaptability [18]. The size difference between autonomous machines and conventional tractors and implements helps mitigate soil-related issues, particularly soil erosion and compaction, brought on by massive, heavy contemporary agricultural equipment [19]. To automate these field activities, an examination of these tasks is required to split them into different robotic jobs. To do this, divide agricultural chores into deterministic (tasks that can be planned and optimized beforehand) and interactive categories [20].

The primary obstacles that agricultural robots must overcome are global in scope and mission-specific. Terrain assessment [21, 22], route planning [23, 24], safety concerns, with a particular emphasis on human detection [25], and fleet robots [26–28] are all global difficulties. The requirements for crop engineering, identifying and categorizing crops or pests, and using inputs are task-specific issues. Most of these problems involve the automation system, the navigation system in semi-structured agricultural areas, the intelligence to govern the automated platform and execution, and the vision system [29].

The apple is the most popular fruit worldwide, and China is one of the major apple-producing nations[1] [30]. Apples are commonly utilized in the culinary business because of their excellent qualities. Juice, jam, marmalade, apple juice, and other goods might be the most significant [31]. The growth of the input industry, which includes the production of weed protection preparations, mineral fertilizer, agricultural mechanization, packaging materials, and the construction of storage facilities, among other things, is greatly influenced by the development of fruit production as a whole, including the production of apples [32–34].

The following industries (input and processing industries) must be developed and improved with the modernization and intensification of apple production to boost the fruit's competitiveness in the global market [35, 36]. The progressive growth of apple production results from the introduction of new breeding methods and more fruitful varieties. This increases the production of higher-quality fruits per unit area and the total apple output [37].

Apple is one of the crops with excellent export potential for farmers and may play a significant role in fostering rural development. It is necessary to evaluate the historical and current conditions of the annual apple production to anticipate future production volumes [38, 39]. It should be parallel to all the characteristics of apple and global production, keeping up with intelligent and modern mechanisms for forecasting the direction of its further development to have a better understanding of the situation in domestic apple production and the likelihood that the improvement of this agricultural branch will lead to global development [40, 41]. So comes the necessity and the utmost responsibility for developing and improving modern mechanisms in various fields of agriculture and its operations, such as harvesting, pruning, thinning, pollination, and bagging of fruits and vegetables, to aid in boosting output and satisfying human requirements.

This study aims to thoroughly analyze the state of the automated technology development of apple fruits and provide a uniform methodology for statistically comparing the most advanced robots in apple fruits and evaluating their commercial viability. The following stages make up the framework for this study: (i) formulating the survey procedure considering the exclusion and inclusion criteria; (ii) searching for and choosing relevant data; (iii) classifying the chosen data; (iv) analyzing and interpreting the chosen data; and (v) reporting findings and conclusions (Figs. 1.1 and 1.2).

The Web of Science database accounted for most of the material chosen for this study. The following keywords are primarily utilized to find relevant scientific papers: Apple's automatic AI system for harvesting, pruning, thinning, pollinating, and bagging. The Web of Science database was searched using the following phrase based on the abovementioned keywords. Of 820 publications, 450 were chosen for the research, and 188 were relevant to our subject (see Fig. 1.3).

This chapter surveys automated strategies and developments for apple fruits on multiple fronts, including harvesting, pruning, thinning, pollination, and bagging (see Fig. 1.1). Based on the key inquiries listed below:

---

[1] https://www.statista.com/statistics/279555/global-top-apple-producing-countries/.

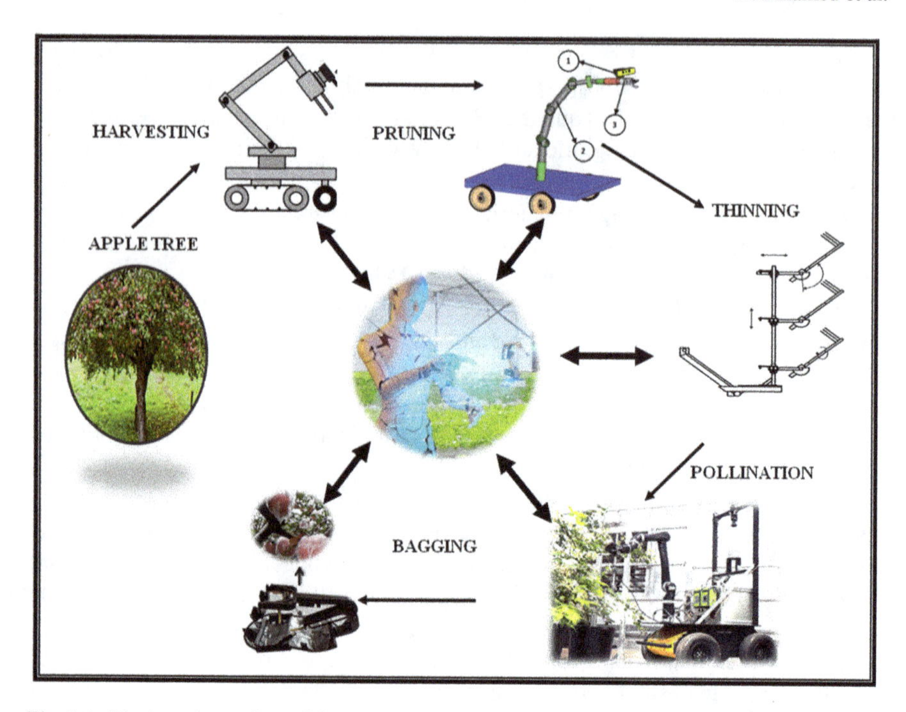

Fig. 1.1 The broad overview of the work

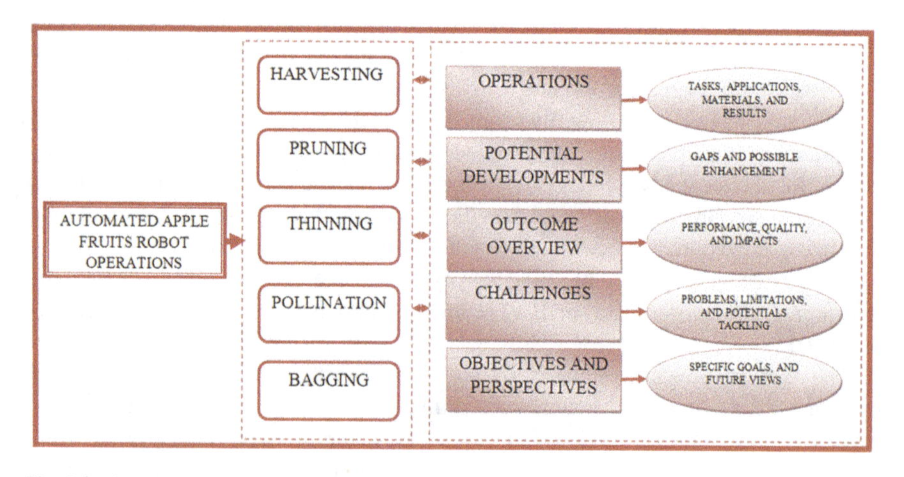

Fig. 1.2 Overall work description

1. What are the components of efficient robots, their benefits and drawbacks, and potential optimizations?
2. Resources, size, accessibility, materials, and data pre-processing?
3. What kind of environment is set up to record performance indicators?

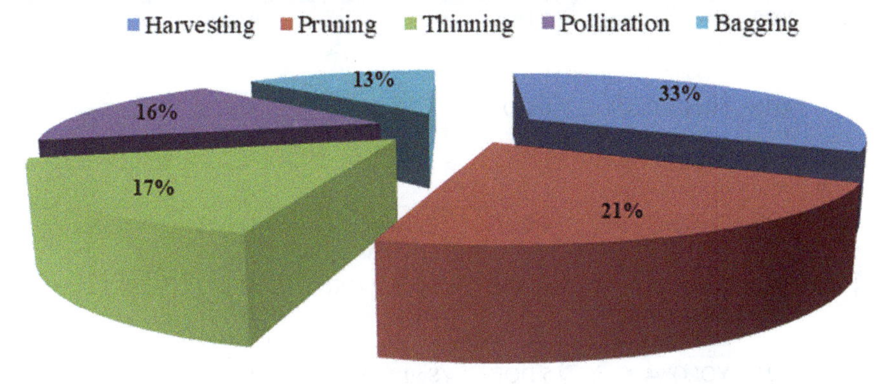

Fig. 1.3 The work's road map description Overall average of the apple fruit operations' work distribution

4. What is the total effectiveness of robots invented over the last decades?
5. What difficulties do the various fields face?
6. Deep learning techniques were implemented, and what were the significant findings?
7. Vision-based machine learning, devices, localization, and identification performance?
8. The key issues and future views?

Then we'll summarize the major issues, the fastest platforms, their efficiency, challenges, and our outlook for the future. The road map of the work is shown in Fig. 1.2.

The rest of this chapter is organized as follows: Sect. 1.2, apple fruit harvesting, contains a brief background, describes the main structure and the components of the system, then summarizes the developments, challenges, and our future views. Section 1.3, apple fruit pruning, includes the automated techniques, abstracts, and our perspectives. As for Sect. 1.4, apple fruit thinning. In Sect. 1.5, apple fruit pollination. Section 1.6, apple fruit bagging. Finally, Sect. 1.7 provides the review's overall conclusions (Table 1.1).

Table 1.1 summarizes the development of the automated apple fruit harvesting equipment

Paper	Year	Hardware	Parameter	Methods	Success rate	Cycle time
[42]	2020	New Catching Arm Apple Picking	7 DOFs	–	93.00%	–
[43]	2020	Mobile-DasNet, and PointNet-based network	6 DOFs	RGB-D	88.00%	6.5 s
[44]	2020	Control System	6 DOFs	Burgers viscoelastic	54.00%	0.48 s
[45]	2021	A hybrid pneumatic/motor actuation system	3 DOFs	Mask R-CNN, GB-D	82.47%	0.3 s
[46]	2021	YOLO v4	5 DOFs	Soft-NMS	92.90%	8.7 s
[47]	2021	A collaborative robot (co-robot) platform	–	–	–	0.015 ms
[48]	2021	LABOR robot	7 DOFs	HSV	72.00%	4.6 s
[49]	2021	Apple harvest's auxiliary tools	–	PID	92.00%	–
[50]	2022	Apple harvesting prototype	4 DOFs	RGB	47.37%	4 s
[51]	2022	Industrial robot KUKA KR 30-3	–	RCVisard 160 vision system	–	–
[52]	2022	A novel three-finger force feedback soft gripper	–	–	96.00%,	–
[53]	2022	Developed robotic apple harvesting prototype	3 DOFs	RGB-D	52.10%	3.6 s
[54]	2022	Dual-manipulator prototype	–	CRITIC–TOPSIS	–	–
[55]	2022	A dual-armed fruit-harvesting robot	6 DOFs	SSD	96.90%	0.5 s
[56]	2022	Apple Retrieving System (MARS)	6 DOFs	Neural Network (NN)	70.80%	7.91S
[57]	2023	A flexible swallowing (FS) gripper	–	CBCM	–	–

(continued)

Table 1.1 (continued)

Paper	Year	Hardware	Parameter	Methods	Success rate	Cycle time
[58]	2023	Soft robotic finger SRF	–	LiDAR-RGB	70.77%	–
[59]	2023	A haptic-enabled robotic gripping technique	–	Deep-touch-CNN	83.3–87.0	–

1.2 Apple Fruits Harvesting

1.2.1 Background

Due to the shortages of human resources and high cost, robotic apple harvesting has drawn much study interest recently. Agricultural professionals have used two different harvesting strategies, selective and bulk, to cut down on labor costs in orchards. Selective harvesting is a robotic technique that uses automatic movements with eventual grasping outcomes. It is often installed on a mobile platform with a final taker and machine vision to pick out certain ripe fruits [60]. It is thought that the mechanical harvesting strategy can potentially replace human gatherers in the long run since automated systems may combine machines' efficiency with people's selectivity [61]. The bulk harvesting technique, which is the second approach to the harvest, relies on the idea that vibrating a fruit tree would make it separate its fruits. Fruit farmers have started using it to produce apples, oranges, and cherries, among other crops [62]. Despite their outstanding efficiency, mass-harvesting devices have considerable limitations [63]. Farmers have expressed worries over excessive damage done to both canopies and fruits by machines [64]. Research on damage reduction in bulk harvesting continues to be active since fruit damage influences their popularity on the market [65]. The quality of the fruits might vary substantially when picked loosely since less-ripe fruits are always taken along with the ripe ones [66]. It is not simple to coordinate fruit ripening rates over an entire orchard. In a bulk harvesting system, the harvest time might rely on reducing losses from picking immature and overripe fruits.

1.2.2 Apple Harvesting Robotics Development

1.2.2.1 Structure of the System's Components

A harvesting robot is often an integrated, interdisciplinary system that combines cutting-edge capabilities from several disciplines, such as mapping, perception, sensing, motion planning, environmental navigation, and auto-mated processing. Consequently, present fruit harvesting robots often include several components: (a) a portable base to transport the robot around the intended object; (b) using machine vision to recognize and perceive the surroundings; (c) a mechanism for controlling the robot generally; (d) a container for collecting and preserving the fruits; (e) more or one manipulator to approach the fruit while avoiding the barriers, and; (f) more or one end-effector may separate the target fruit from the plant.

Robots in Traptic[2] and FFRobotics[3] use conveyors to stop fruit damage while being collected, while Bac et al. [67, 68], Lehnert et al. [69–71], Boaz Arad et al. [72] possess vertical lifting tools to increase the robot's work area.

1.2.2.2 Robots Technology for Apple Harvesting

Apple harvesting is one of the crucial operations, and the employment of modern tools and techniques has expanded to improve quality and boost output. Over the last several decades, intelligent fruit-harvesting robots have been actively developed to fill the widening gap between feeding a population that is expanding quickly and the human resources available. Table 1.1 shows how apple fruit harvesting techniques have developed, including the dates, activities, application strategies, and outcomes. This section will start with hardware (see Fig. 1.6) and later focus on machine vision strategies and recognition systems (Fig. 1.7).

(A) Hardware

To evaluate the performance of this novel catching arm design, a virtual apple field of 505 apples was employed [42] to test the picking capabilities of a 7 DOF arm. For the virtual apple field that was made available, with a maximum drop height of 30 cm, the goal-capturing efficiency was 90%. The design was finished, manufactured, and verified using the sophisticated mechanical linkage design. According to the workspace study, the efficiency climbed to an acceptable 93% when the drop height was raised. Kang et al. [43] put out Mobile-DasNet, a lightweight, one-stage network with high computational efficiency, to conduct fruit identifications and instance segmentation based on sensory input. They suggest a modified PointNet-based network to represent fruits and estimate their reaching distance using RGB-D camera point clouds. They incorporate the above two characteristics into developing a precise robotic system for autonomous apple harvesting (Fig. 1.4a). Regarding F1 score, recall, and accuracy for fruit identification, the suggested Mobile-DasNet obtained 0.851, 0.826 and 0.9, and an accuracy of 0.82 for instance segmentation. The grasping estimate, meanwhile. In the orchard scenario, the IoU3Ds obtained by the PointNet grasping estimate, RANSAC, and HT algorithms were 0.88, 0.76, and 0.78, respectively. The designed robotic harvesting system has a cycle duration of 6.5 s and a harvesting success rate of 0.8 overall. In [44], the harvesting robot experiment setup and the control system simulation model are established (Fig. 1.4b). The efficiency of the suggested strategy is shown by comparing simulation results and

[2] https://www.traptic.com/

[3] https://www.ffrobotics.com/

conducting grasping tests on a harvesting robot. According to simulation and experiment data, the proposed control is smoother than the intended force, and its overshoot is roughly 2.3%. The reaction time is quicker, and the contact force adjustment time is reduced by around 0.48 s. Compared to the conventional force-based impedance control, the contact force overshoot is roughly 2%, which is 37.5% less.

They created an RGB-D camera-based fruit localization and identification system based on deep learning [45]. A three-degree-of-freedom manipulator (Fig. 1.4c) uses a hybrid pneumatic/motor actuation system to enable dexterous motions; the apple is detached using a vacuum-based end-effector. 3 DOF manipulator, RGB-D camera, and a vacuum-based end-effector main components of the system design, then trained on the dataset, split into training, validation, and test, 10,530, 4203, and 4795, respectively. Mask R-CNN accuracy up to 90.5%. According to field testing, 80 of the 97 fruit attempts were successfully separated from the tree. The best average time was just 0.3 s, and the picking efficiency was 82.47%.

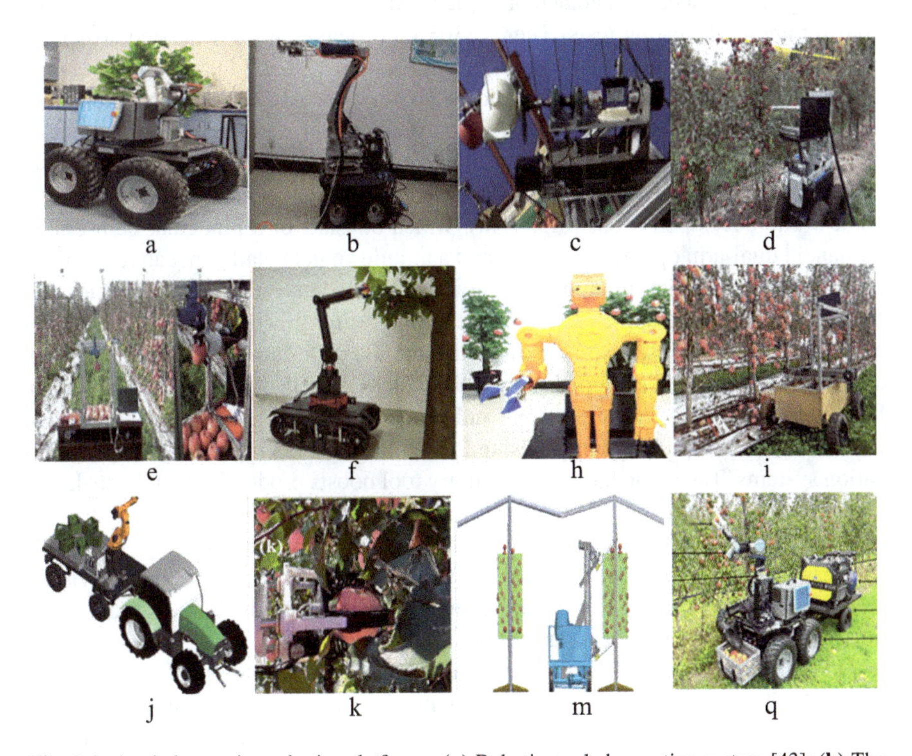

Fig. 1.4 Apple harvesting robotics platforms: (**a**) Robotic apple harvesting system [43]; (**b**) The apple-picking robot's trial base for gripping [44]; (**c**) Presents the structure of the experimental device [45]; (**d**) The developed robotic apple harvesting prototype [73]; (**e**) Apple picking grasp force device [74]; (**f**) Self-made apple-picking automaton [75]; (**h**) LABOR harvesting robot [48]; (**i**) Apple harvesting robot prototype [50]; (**j**) Automated fruit-collection device graphic illustration [51]; (**k**) Soft fingers [76]; (**m**) Robotic harvesting optimization using two manipulators in a simulative field [54]; (**q**) Apple retrieving system at Monash [56]

A Des-YOLO v4 algorithm and an apple identification approach are suggested [46] to enable harvesting robots to rapidly and correctly identify apples in complicated environments. After that, 10,100 Apple photos were examined, and the results were compared to the most common detection methods. A class loss function based on Average Precision Loss (AP-Loss) is presented to address the abovementioned issue and increase the accuracy of apple identification. With a mean average precision of 97.13% for detecting apples, a recall rate of 90.00%, and a detection speed of 51 f/s, the Des-YOLO v4 network boasts excellent characteristics. According to the testing, the robot's harvests are completed 92.90% of the time, taking 8.7 s. Fei et al. [47] converted a common harvesting platform into a platform for collaboration robots (co-robots) (Fig. 1.5g). The co-robotic platform moves ahead, employing a vision system to predict the distribution of the incoming fruit, instrumented picking bags to gauge each worker's picking pace, and hydraulic lifts to raise and lower employees as needed. The model-based control algorithm increases the machine's harvesting efficiency and reduces the damage of apple picking. A commercial orchard was also used for apple harvesting tests. There, 2307 kg of apples were harvested, 1045 kg in a zone harvesting mode with variable heights, and 1262 kg in a zone harvesting mode with fixed measurements specified by the producer. At a human-controlled platform moving speed, the throughput of the variable-height zone harvesting mode was 327.6 kg/h compared to 298.8 kg/h for the fixed zone harvesting mode, an increase of 9.5%. They offer LABOR [48], an autonomous humanoid robot, for picking apples. The developed robot (Fig. 1.4h), which consists of a binocular cam-era, a humanoid dual-arm operating system, and a mobile vehicle platform, can recognize, position, grip, and pick up apples using its dual-arm harvesting and vision identification systems. The apple recognition and harvesting functionalities had 82.50% and 72% success rates, respectively, while the average duration for gathering an apple was 14.6 s. The developed robot [49], which consists of a binocular camera, a humanoid dual-arm operating system, and a mobile vehicle platform, can recognize, position, grip, and pick up apples using its dual-arm harvesting and vision identification systems. The apple harvest's auxiliary tool boosts productivity and cuts labor costs. And the collecting efficiency was 4900 apples/h, 5300 apples/h, and 5700 apples/h when the motor speed was 52, 56, and 60 r/min, respectively, while the fruit damage rates were 4%, 5%, and 8%. Compared to human operation and visual recognition-based picking robots, the findings showed that this device's production efficiency was rising without causing increased fruit damage.

Hu et al. [50] created a 4-DOF manipulator-based robotic apple-picking prototype (Fig. 1.4i). The vacuum end effector, 4-DOF manipulator, and binocular camera were all built into the picking prototype. It comprises a self-built Ackerman steering mobile vehicle, a vacuum-based end-effector, a binocular camera (ZED 2, Stereolabs Inc.), and a 4-DOF manipulator. During use, the manipulator is put together on one side of the row of fruit trees, and the vacuum-based end-effector is attached to its end. Also, it compared the report's performance to the other apple harvest robotics. With a picking cycle length of 4 s, the field orchard's picking success rate was 47.37%, while the simulated orchards were 78%. In the field orchard, 11.11% of the stems were damaged.

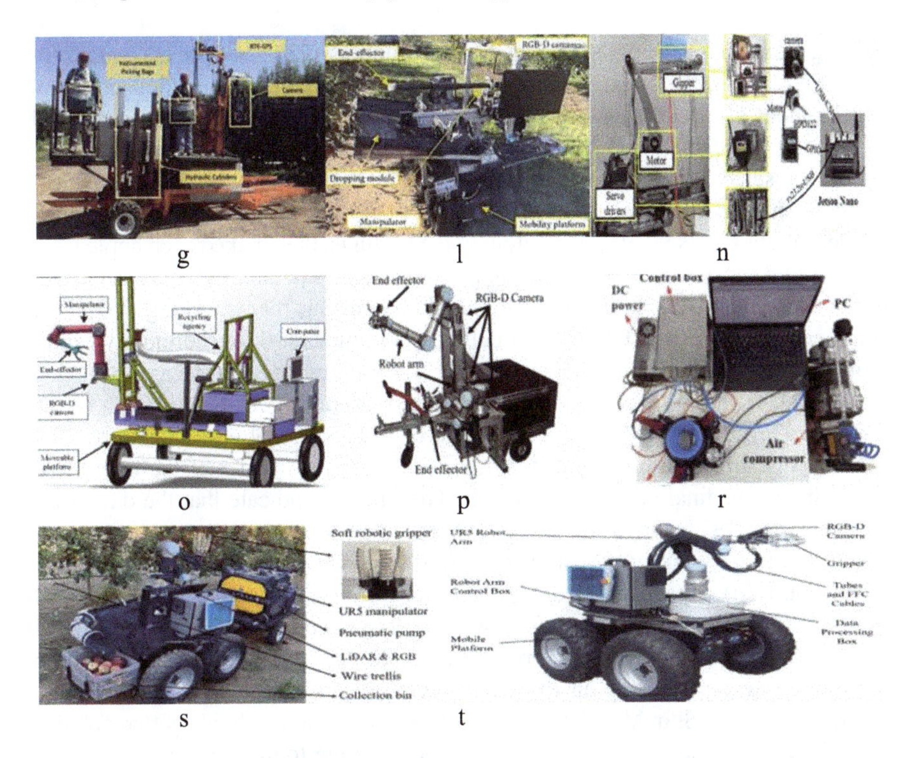

Fig. 1.5 Robotic platforms for picking apples: (**g**) Robotized structure with employees [47]; (**i**) Apple harvesting robot prototype [50]; (**n**) The trial device for apple harvesting [77]; (**o**) An automated fruit-harvesting device [79]; (**p**) Dual-arm fruit picking automaton automated gathering [55]; (**r**) The FS gripper's base for gripping [57]; (**s**) Apple picking technique at Monash [58]; (**t**) The combined apple picking equipment [59]

An automated system for harvesting apple crops is being developed as part of the robotic complex [51]. It is positioned on a moveable platform with wheels (Fig. 1.4j). An industrial robot, a packing system, a vision system, a control system, a vacuum pump, an electric generator, and a container comprise the sophisticated machinery on the mobile platform. The platform travels across the garden between the trees and a tractor. By attaching the electric generator to the tractor's power take-off shaft, the issue of autonomous power supply for electrical equipment is resolved. In [52], they offer a soft gripper for apple picking (Fig. 1.4k). To lessen the chance of apple skins being harmed during mechanical harvesting. A novel three-finger force feedback soft gripper is suggested for the apple harvesting robot. Hard fingerpicking has a 100% success rate but a 16% harm rate. In comparison, the success rate of soft finger picking with slip detection climbed to 96%, while the damage rate reached 8%. Soft fingerpicking without slip detection had an 80% success rate but caused no harm to the fruit peel.

Kaixiang Zhang et al. [53] discussed the essential methods and advancements in multi-view fruit localization and identification using deep learning, coordinated picking, dexterous manipulation control, and dropping planning. The current prototype (Fig. 1.5l) demonstrates encouraging performance in the direction of future development of effective and automated apple harvesting technologies. A total of 142 apples were evaluated in this outdoor test, and 74 were effectively selected, yielding a selection percentage of 52.1%. The recognition algorithm with a two-camera configuration receives an assessment of 93.92%, while the technique with a single camera scores 90.5%. According to the findings, they picked each apple in an average of 3.6 s. With an apple-picking pace of 7 to 10 s per apple, this is a substantial advancement over earlier recorded apple-harvesting robots. Xiong et al. [54] combined the CRITIC-TOPSIS technique as the foundation for thoroughly examining the dual manipulator arrangement (Fig. 1.4m). The length of the upper picking arm is 1119.3 mm, the angle of the horizontal elevation is 39.4, the length of the lower picking arm is 898.7 mm. The angle of the horizontal elevation is 26. The base for installing the outer frame is 755.3 mm from the center of the tree trunk, according to the final optimization result. The findings indicate that the dual-ideal manipulator's design may completely cover the target working area, and its duplication rate is 16.62%. Takeshi Yoshida et al. [55] created a dual-armed fruit harvesting robot (Fig. 1.5p) to reach most fruits on a specially grown and adjusted joint V-shaped trellis. The fruit harvesting robot employs sensors and computer vision to locate the fruit and estimate its location before inserting end-effectors into the fruit's bottom portion to harvest it. Using an RGB-D camera, they locate the fruits in a picture and use the Single Shot Multibox Detector (SSD) detection techniques to recognize fruits in photos. They demonstrated that route planning to the harvesting objective could be completed relatively quickly in less than 0.5 s using inverse kinematics and T-RRT. The highest performance was 96.9%, which improved the apple harvest Robot's efficiency and quality.

Wesley Au et al. [56] presented the Monash Apple Retrieving System (MARS) (Fig. 1.4q), a platform for selective apple harvesting that can locate and collect apples in situations with complicated canopies without the need for canopy simplification. In-depth tests were conducted in Australia during the 2021 and 2022 apple harvesting seasons. MARS successfully gathered 62.8% of the apples with a cycle duration of 9.18 s and minor fruit damage. Under optimal visibility, performance improved to a success rate of 70.8% and a cycle time of 7.91 s. In [57], an FS gripper was designed (Fig. 1.5r) and developed first, followed by a force-to-deformation model of the FRE finger based on CBCM, a grasping force sensing model related to the FRE finger's bending angle to perceive grasping force in real-time, and finally, grabbing experiments to test the gripper's force sensing accuracy and adaptability. The findings showed that the mean absolute error was 0.153 N, and the average relative error was 5.65%. The maximum bending angle and contact force were 22.6 and 5.72 N, respectively; better ability and more considerable tensile strength were achieved when grasping the finger's base. The unique soft robotic gripper proposed in [58] has suction cups with active motion, passive compliance, and tapered delicate robotic fingers (Fig. 1.5s). The tapered SRF is a single-step 3D printed customized, flexible bending actuator. The suggested robotic gripper is small, supports apple grabbing, and produces significant grasping force. The harvesting rates were 75.6%, 4.55%, and 70.77%, respectively.

Finally, we wrap up the development of apple fruit harvesting instruments by [59], a haptic-enabled robotic gripping technique that was suggested to integrate soft robots, touch sensors, and deep learning. The robot (Fig. 1.5t) can detect and manage branch disturbances during the gathering process, thereby decreasing possible mechanical fruit harm, which was accomplished by combining fin-ray fingertips with integrated haptic sensor arrays and unique awareness techniques. The outcome demonstrates a total success rate of 83.3–87.0% in detecting the gripping state, proving the potential interference management technique and harvesting quality.

Some difficulties remain, but creating high-throughput phenotyping ma-chines for automated movement may overcome them. The height of the same product at various development phases is secondary due to the diversity of crops, and the farming method is distinct. Some phenotyping robots have a set height and wheel spacing, which prevents them from being used in other contexts and allows them to be applied only to a particular crop or farming scenario. Second, installing numerous instruments and sophisticated control systems drives the price of constructing a robot platform. A significant problem is figuring out how to lower machine-making expenses. Third, the robot's proper working area is constrained to an ample space of field products and is only a few square meters in a motionless condition.

Furthermore, a hectare of crops' trait data may need to be collected over several hours due to the complex field environment's limitations on moving pace. Fourth, real-time decision-making is not possible because most data processing and phenotypic parameter extraction occur after the phenotypic platform has been collected. Consequently, the addition of real-time data processing devices is required.

The apple machines performed admirably in (Figs. 1.4 and 1.5), cutting labor requirements and time, luring young engineers and farmers into the agricultural sector business, and raising output. First, we note that [42, 53, 53, 45, 47–59] worked with high efficiency on different tasks, but we observe that [55] was the highest harvesting performance with a rate of up to (96.90%). Considering the cost of the design and development processes, we see that the growth in the value of the parameters impacts the speed and success rate.

As is widely known, robots are costly [59]. It takes a while for development, so it's necessary to keep improving Apple's automated equipment and components, such as arm selection, manipulators, artificial digits, etc., which help to enhance quality, provide practical tools, and require little maintenance (Table 1.2).

(B) Vision strategies and recognition systems

Robots employing vision systems can adapt efficiently. Due to its versatility, the assembly process may go faster without being slowed down by erroneous placements. Provide tighter quality control. When used to evaluate equipment, spot flaws, increase speed, or develop advanced robotics [93]. Table 1.2 shows the development of vision techniques and detection systems, including the dates, features, approaches, datasets, algorithms, and performance.

Table 1.2 The development of vision techniques and detection systems

Paper	Year	Features	Data size	Algorithms	Result
[78]	2020	Color and shape	30 images	Adaptive Gamma correction	84.00%
[80]	2020	Defect	1,386 images	Fast-FDM	62.30%
[81]	2020	Color and size	12,800 images	SNAP system using Faster R-CNN	87.90%
[82]	2020	Color and shape	528 images	Suppression Mask RCNN	90.50%
[83]	2020	Detect, count, and size	OIDv4 datasets	YOLOv3-tiny	83.64%
[84]	2020	Color, shape, and texture	1899 images	U-Net segmentation	97.91%
[85]	2020	Color, texture, 3D shape	400–800 images	Deep neural network DaSNet-v2	79.40%
[86]	2020	Color, shape, and texture	878 image	YOLOv3 algorithm	92.8%
[87]	2020	Color, size, and detect	800 images	VGG16	89.30%
[88]	2020	Color, shape and Defect	150 images	PCNN and GA	94.88%
[89]	2020	Color, space, and Defect	150 images	K-means	–
[90]	2020	Color, shape, and texture	1020 image	Mask R-CNN	86.14%
[91]	2021	Color, shape and texture	878 images	YOLOv5 algorithm	97.80%
[46]	2021	Color and shape	10,100 images	A Des-YOLO v4	93.10%
[75]	2021	Color, shape, and texture	2670 image	EficientNet-B0-YOLOv4	96.54%
[92]	2022	Color and shape	17, 930 images	Shufflenetv2-YOLOX	96.76%
[77]	2022	Color and size	–	YOLOv3, RANSAC	96.00

In [78], they presented a technique of picture correction based on the adaptive Gamma approach. After combining, the fundamental image processing techniques of corrosion, expansion, whole filling, and fixed threshold segmentation were used to create the treatment process to achieve a clean fruit region in the amassed apple picture. Next, using an efficient, iterative, open procedure, a near-large fruit was generated from an apple picture of an orchard. Lastly, apple photographs from the collected images were chosen following substantial to low light conditions to test the suggested procedure. With an average and maximum reduction ratio of 70 and 84%, respectively, the improved iterative, open operation decreased the algorithm's execution time. Jia et al. [80] provided a quick and precise object identification model, called Fast-FDM, for fast real-time green apple detection by harvesting robots, utilizing the extensive green apple dataset gathered in complicated orchard environments. Use a small-sized EficientNetV2 with a BiFPN backbone network to extract and fuse features, increasing training speed with fewer parameters and FLOPs. Last, they use the ATSS approach to adaptively choose training samples, improving the recall of targets at various scales and offering additional opportunities for efficiency development. In [81], a faster region-convolutional neural network technique was implemented for multi-class apple recognition in dense-foliage fruiting wall trees. Eight hundred photos were captured and subsequently enhanced to create 12,800 images. Fruit with no blocks, leaves, stems, or wires and fruit with obstacles all had average precisions of 0.909, 0.899, 0.858, and 0.848, respectively. The average processing time for a picture was 0.241 s, and the mean average accuracy of the four classes was 0.879 overall. Pengyu Chu et al. [82] created a new deep network called the suppression Mask RCNN. Extensive tests demonstrate that the suppression Mask R-CNN network beats state-of-the-art models with a higher F1-score of 0.905 and a detection time of 0.25 s per frame on a typical desktop machine.

For real-time apple identification [83], they used an altered version of the YOLOv3-tiny algorithm on embedded systems like the Raspberry Pi 3 B+ in conjunction with Intel Movidius Neural Computing Stick (NCS), Nvidia's Jetson Nano, and Jetson AGX Xavier. The average detection accuracy was 83.64% in the findings, and a frame rate of 30 fps was attained. Li et al. [84] proposed an ensemble U-Net segmentation model suitable for small sample datasets containing 211 apple images, the augmentation task on it, the highest performance up to 98.32% with a recognition speed of 0.39. By expanding the application domain of harvesting robots and orchard yield measures, the suggested technique offers a theoretical benchmark for additional target fruit segmentation initiatives in the apple.

An advanced neural network, DaSNet-v1, was created in [85] for segmenting and detecting fruits and branches in an orchard setting. Experimental data collected during field experiments in an apple orchard were used to evaluate and verify DaSNet-v2. Resnet-101 and DaSNet-v2 combined yielded results of 0.868 and 0.88 on recall and precision of detection, 0.873 on the accuracy of segmenting fruits, and 0.794 on the accuracy of segmenting branches, respectively. DaSNet-v2 with ResNet-18 achieved recall and precision of detection at 0.85 and 0.87, fruit segmentation accuracy at 0.866, and branch segmentation accuracy at 0.757, respectively. In [86], they developed an automated technique for spotting apples in orchards. The system was created for harvesting robots based on the YOLOv3 algorithm with unique pre- and post-processing. The suggested pre- and post-processing methods allowed the YOLOv3 algorithm to be modified for use in an apple-harvesting robot machine vision system, resulting in an average apple detection time of 19 ms with a share of objects misidentified as apples at 7.8% and a share of unidentified apples at 9.2%. Longsheng Fu et al. [87] used a low-cost RGB-D camera to create an outdoor machine vision system to better recognize Scifresh apples in fruiting-wall trees by removing background objects using depth characteristics. The Zeller Fergus Net (ZFNet(and Visual Geometry Group with 16 layers (VGG16) architectures, which are the two most popular Faster ReCNN models, were modified and put into practice.

The study found that the Foreground-RGB pictures using VGG16, which on average took 0.181 s to analyze a 1920 × 1080 image, obtained the most excellent average precision (AP) of 0.893.

On extracting features in [88], they use PCNN models for their apple datasets. The 16 retrieved feature vectors effectively describe the object's global and local characteristics. The Elman neural network is then tuned using a genetic algorithm, which increases the network's capacity for generalization and sharply raises the training success rate, which in this research reaches 100%. It demonstrates how the new GA-Elman model enhances net-work operating effectiveness and provides the most significant recognition rate. The apple picture is first converted to the Lab color space [89], then the K-means method is used to segment it. Second, morphological processes, including erosion and dilation, abstract the apples' contour. Then, picture points are separated into the core, edge, and outside issues. Third, every internal point's minimal distance from the edge is determined using a quick method. Then, by locating the maxima among these distances, the apple centers are obtained. The last step is calculating the radii by determining the shortest distance between the center and the edge. Positioning is thus accomplished.

The model [90] was enhanced to identify better and segment apples with overlaps. As a backbone network for feature extraction, Residual Networks (ResNet), in combination with Densely Connected Convolutional Networks (DenseNet), may significantly decrease input parameters. The region of interest (RoI) is created using feature maps as input to the Region Proposal Network (RPN) for end-to-end training, and the apple's location is determined using the mask made by the full convolution network (FCN). A random test set of 120 photos is used to evaluate the approach, and the results show that the precision rate has achieved 97.31% and the recall rate has reached 95.70%. Additionally, the identification speed is quicker and can match the demands of the vision system of the apple harvesting robot. YOLOv3 and YOLOv5 apple-harvesting robots algorithms for apple identification were tested and evaluated [91] in 878 images of apples of various kinds. Outcomes show that the YOLOv5 algorithm could identify apples in orchards with 97.8% recall (fruit detection rate) and 3.5% False Positive Rate without extra pre- and post-processing (FPR). It performs much better than YOLOv3, which provides 9.1% recall and 10.0% FPR without pre-and post-processing, 90.8% Recall, and 7.8 FPR when paired with specific pre and post-processing techniques.

In [75], the relevant apple photographs from the Internet are gathered for classification using crawler technology; after that, a leaf data augmentation method was developed to address the issue of inadequate picture data brought on by the random occlusion between leaves. They proposed the EficientNet-B0-YOLOv4 model, compared with OLOv3, YOLOv4, and Faster R-CNN with ResNet, on 2670 expanded samples. Their method performed best; recall, precision, and F1 average values are 97.43%, 95.52%, and 96.54%, respectively, while the model's average detection time per frame is 0.338 s. Wei Ji et al. [92] suggested a Shuflenetv2-based YOLOX technique for detecting apples (Fig. 1.5n). The lightweight network Shuf-flenetv2 combined with the convolutional block attention module (CBAM) serves as the framework for this technique, which employs YOLOX-Tiny as the baseline. The PANet network is enhanced with an adaptive spatial feature fusion (ASFF) module to increase detection precision, and the network structure is made simpler by using only two extraction layers. The results show that Shuflenetv2-YOLOX's average precision (AP), precision, recall, and F1 are 96.76%, 95.62%, 93.75%, and 0.95, respectively. Compared to the advanced lightweight networks YOLOv5-s, Eficientdet-d0, YOLOv4-Tiny, and Mobilenet-YOLOv4-Lite, it has a more significant detection effect and speed. In [77], they provide a technique for using the point cloud of the apple and the surrounding area to determine the best picking orientation concerning the target fruit and nearby branches (Fig. 1.5o). A point cloud of the apple and the area around it is then produced once the YOLOv3 target identification algorithm first recognizes the apple. The fruit is idealized as a sphere for sphere fitting, and the random sample consensus procedure RANSAC is applied. The standard deviation was 13.65, and the average angle error was 11.81 between the estimated picking direction vector and the predicted direction vector. Of the determinations, 62.658% were off by 10, and 85.021% were off by 20. An apple's orientation was estimated at an average of 0.543 s.

When we get to the second part, which deals with the robot vision system, we see that the most popular methods were CNN with various types such as (mask R-CNN, DaSNet v2, VGG16, Fast FDM, DaSNet v2, Fast R-CNN, Shufflenet), YOLO v5, 4, 3 algorithms, GA-Elman, SIFI, and SSD. Fast R-CNN was the most widely used, but the YOLOv5 algorithm had the best speed and an accuracy of 97.80%. Before being installed to any company's tasks, developing approaches and algorithms requires complex procedures, intensive processes, and extensive training and testing.

Regarding the apple fruit robotic system, the detection accuracy directly impacts the performance. As a result, it is critical to develop new methods, add custom functions and processes, and stay up to date with technological advancements by using the transformer approach to improve recognition and handle large amounts of data. These factors affect the robotic system's speed, accuracy, quality, and output of apple products. We conclude with the resources and materials of the apple items by utilizing the advanced collection technology and enhancing its innovative preprocessing methods, plus extending the free availability of the datasets to the research community.

1.3 Apple Fruits Pruning

1.3.1 Automated Pruning

Pruning involves cutting tree branches for several horticultural and financial reasons, including plant reproductive growth and balancing vegetative growth; limiting plant size; modifying the canopy to affect fruit quality, size, and yield; and determining the best crop load for the following season [79]. Because fruit trees are perennials, pruning is a cumulative process, impacting plant growth and productivity for ages afterward [94].

As described in the literature to date [95–99], instrumental pruning entails recognizing the geometry of the tree, putting a pruning procedure into action, using automated navigation, and chopping branches. To estimate an object's form with high accuracy, well-established computer vision algorithms often depend on objects having distinguishable characteristics, such as angles or edges, a stable or predictable picture collection environment, or an object's smoothness [100]. Trimming occurs when the trees are dormant, the object is thin, and information is gathered from the outside. These characteristics break several presumptions of traditional approaches [101]. These factors have led to development of novel perception techniques for tree canopies. Lastly, visual access to the tree canopy is limited by the distance between trees and permanent trellis constructions, and photos are often only captured from one side of the tree at a time (Figs. 1.6 and 1.7).

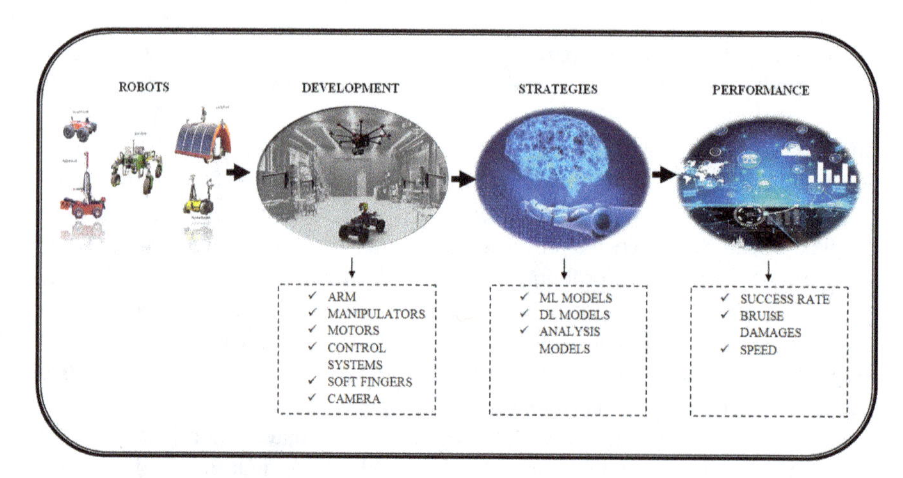

Fig. 1.6 Roadmap for analysis of developing the equipment

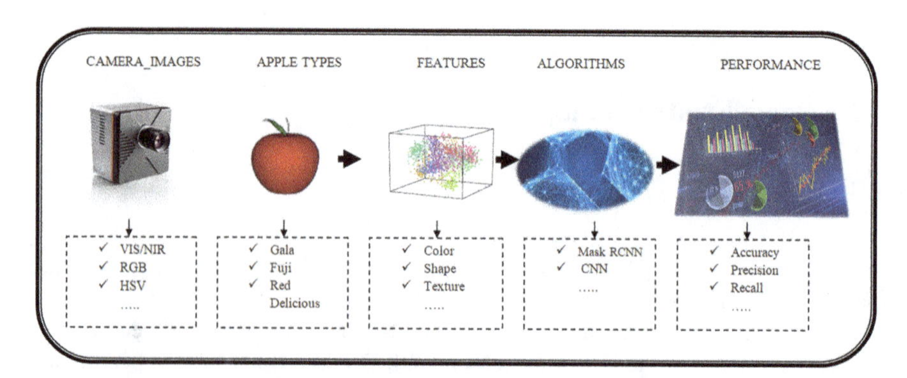

Fig. 1.7 Overall description of development for the vision systems and recognition approaches

1.3.2 Apple's Robotic Pruning System's Key Components

Robotic pruning, which makes precise cuts using a cutting tool coupled with a robotic arm, is a particular branch operation [102, 103]. Generally, a robotic pruner consists of a vision sensor, a mechanical manipulator, and an end-effector tool (Fig. 1.8).

1.3.2.1 Machine Vision System

One of the essential elements in creating a robotic pruner for tree fruit is the ability to recognize branches. Effective barrier 'branch' identification is also a need for route design. In the last ten years, a broad range of tree fruit crops has been explored using machine vision, a promising branch identification technique [104–106]. One significant benefit of adopting machine vision methods is their ability to provide non-destructive, accurate, dependable, economical, and automated solutions for orchard management operations [107].

(a) Branch detection for apple pruning

The process of identifying the undesirable branches that need to be pruned is known as branched detection. Color, stereo vision, and time-of-flight cameras are just a few machine vision sensors that identify unproductive shoots and branches without causing damage [108]. Segmentation is a frequently used approach in object identification to divide an image into different parts depending on the properties of the picture's pixels. It entails classifying backdrop pixels such as foliage, fruits, and branches. Various branch segmentation techniques, such as deep and machine learning, have been tested on fruit farms [109]. Table 1.3 summarizes the apple branch detection's dates, features, cameras, techniques, and performance.

Karkee et al. [104] created 3D skeletons of tall spindle apple trees using a time-of-flight-based 3D camera and thinning algorithms to identify pruning branches. Ji et al. [105] evaluated contrast-limited adaptive histogram equalization (CLAHE) for segmenting apple tree branches using the RGB camera and reported a high

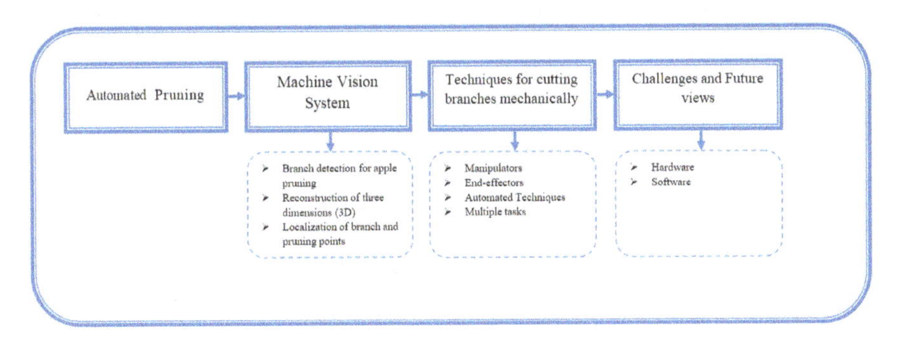

Fig. 1.8 Overviews of automated apple pruning developments

Table 1.3 An overview of the various characteristics and techniques for sensors used in apple branch detection

Paper	Year	Sensors	Features	Segmentation approach	Performance
[104]	2014	3D CamCube 3.0 time of flight camera	3D color images	Skeleton analysis by medial axis thinning algorithm	Branches recognition accuracy (77.00%)
[105]	2016	RGB Camera	Color image	Contrast limited adaptive histogram equalization (CLAHE) based segmentation	Detection accuracy: 94.00%
[106]	2018	Kinect v2	RGB, depth, and pseudo-color images	R-CNN with AlexNet network	Accuracy: 85.5%, and Average recall rate: 91.50%
[110]	2020	Kinect v2	RGB and point cloud data	CNN-based segmentation (SegNet)	RGB and Foreground-RGB accuracy was 89.00%, and 92.00%, respectively

identification rate. They concluded that the CLAHE-based segmentation approach outperformed the OTSU and histogram methods for apple branch segmentation.

3D vision camera techniques have significantly enhanced the effectiveness of branch identification. Three pictures, comprising RGB, depth, and index images, were captured using a stereo vision camera to identify apple branches, and the mean accuracy was increased by Zhang et al. [106]. A Kinect v2 camera was employed by Majeed et al. [110] to identify apple tree branches. A CNN-based SegNet segmentation network separated the apple tree's branches using foreground RGB pictures, and the highest performance was up to 92.00%. The Cutout, CutMix, Mixup, SnapMix, and Mosaic algorithms are used in [111] for data augmentation and other picture preparation techniques. The pruning inference was developed to address the issue of slowing down the training and inference due to the increasing complexity of detection networks. The suggested model has an inference speed of 29 FPS and can recognize apple blooms with accuracy, recall, and mAP of 90.01%, 98.79%, and 97.43%, respectively.

(b) Reconstruction of an apple tree in three dimensions (3D)

Making judgments about pruning requires knowledge of the diameter, orientation, structure, and appearance of the trunk and branches, all of which may be obtained through three-dimensional (3D) reconstructions of trees [112].

In [113], they created and verified a 3D model of an apple tree during its dormant season (Fig. 1.9p1). Next, an algorithm was devised to identify pruning branches and the technique of identifying pruning branches was validated using accurate human pruning branch recognition. Branch identification and pruning branch identification algorithms have 76% and 86% accuracy rates, respectively.

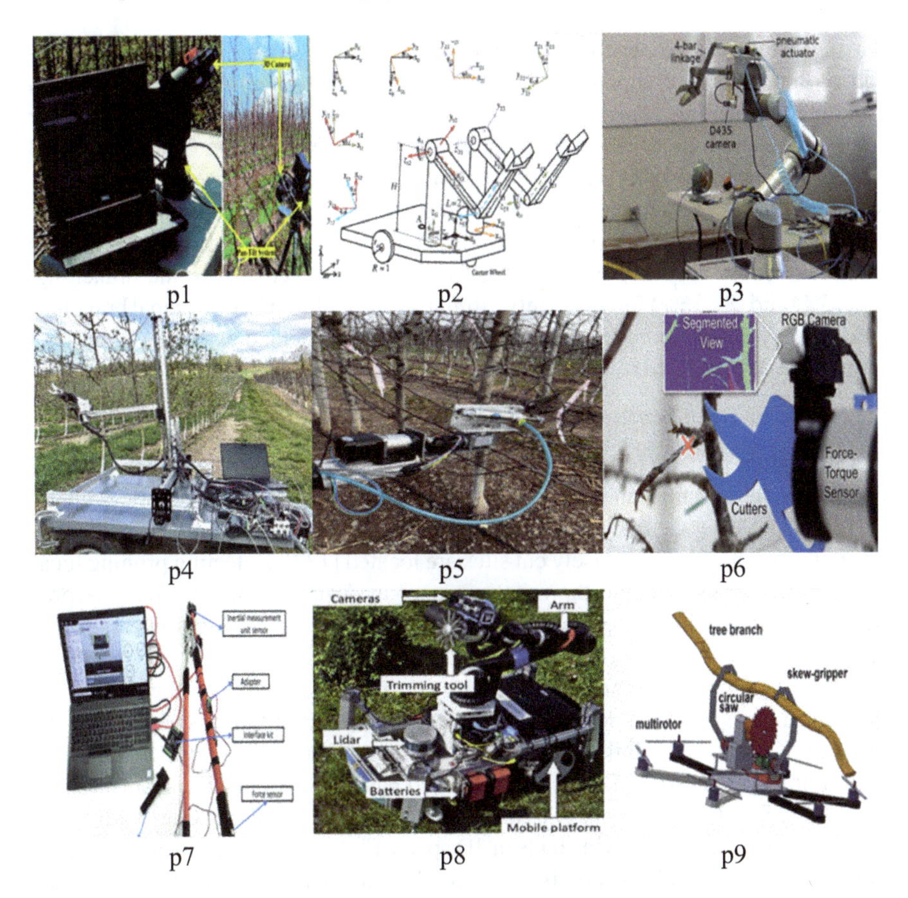

Fig. 1.9 Apple pruning robotics platforms: (p1) Identification platform for apple pruning [114]; (p2) Symmetrical design, dual-arm mobile robotic manipulator [112]; (p3) Pruning robot platform [115]; (p4) A technique integrating pruning [116]; (p5) The developed robotic pruning equipment [117]; (p6) Hybrid vision approach for automatic pruning device [118]; (p7) Apple's robot pruner platform [119]; (p8) Robot for trimming boxwood; (p9) Robotic Flying Trimmer

Manoj Karkee et al. [120] outlines a technique for obtaining apple trees' 3D structures and identifying branches to create an automated pruning system. The apple trees in a commercial orchard are captured in three dimensions using a time-of-flight-of-light-based camera (ToF 3D). Their technique correctly detected trunks while reconstructing the 3D architecture of apple trees. The approach has a false negative identification rate of 23% and a branch identification accuracy of 77%. In a 3D reconstruction approach, the role of representing a tree's trunk and main branches using semicircles is examined in [121]. There are three primary processes: calculating the diameter error, filtering the depth pictures, and semi-circle-based modeling, which estimates the necessary pruning data.

Moreover, they put forward a brand-new empirical model that calculates the diameter of the main branches given a range of depth values. The findings demonstrate that the suggested scheme per-forms at an accuracy of 89% when predicting the diameter of the principal components. In [67], they created a reliable 3D reconstruction method for modeling a dormant apple tree's trunk and principal branches using color information and time-of-flight depth data from the Kinect2 sensor. Quantitative and qualitative comparisons of their suggested method with a depth-based reconstructing strategy show the value of using color information in our automated pruning system. Within error margins of 3 mm and 5 mm, respectively, the suggested method performs with an average accuracy of (93.94%) for adequately recognizing the branching, (71.13%) and (89.26%) for accurately estimating the widths of the principal branches. Additionally, the temporal complexity of our suggested method exhibits a significant improvement compared to the complexity of the baseline strategy.

(c) Localization of branch and pruning points

Finding the pruning points, which direct the cutter to undertake branch removal operations on unwanted and unhealthy branches, is essential for the robotic pruner's performance. The accuracy of identification and the environment in which trees develop determine how precisely cut sites are located [122]. Detecting pruning locations is still tricky, keeping the robotic pruner in development. In a rebuilt apple tree, a few efforts have been made to identify pruning spots using a two-step rule [104, 123].

Pruning guidelines were established with a cut point 0.2 m out from the trunk along the branch line and thresholds for the branch length and interbranch spacing (0.2 and 0.08 m). Karkee and Adhikari [123] found an overall accuracy of roughly 90% in identifying pruning spots. Dáz et al. [124] used Fast Approximate Nearest Neighbor (FLANN), SVM, and Density-Based Spatial Clustering of Applications with Noise (DBSCAN) to locate grapevine buds in 3D space. FLANN was used to match critical points, SVM to classify 3D points, and DBSCAN to locate 3D buds. The placement error was within 1.5 cm, equivalent to around three bud diameters. Katyara et al. [125] found a remarkable trimming performance of 95.1% while moving randomly in each predetermined orientation. The accuracy for the stationary, y-translation, x-rotation, and z-rotation, respectively, was 100%, 97.2%, 98.7%, and 96.4%, demonstrating the equipment's vision system's capability to perform robotic trimming in actual circumstances.

1.3.2.2 Techniques for Cutting Branches Mechanically

(a) Manipulators

DoFs and joint types (revolute or prismatic). Each joint in the manipulator has one degree of freedom (DoF) and standard configurations, such as total DoFs. The type of joint significantly impacts the kinematic dexterity, obstacle avoidance skills, and spatial requirements during manipulation to attain different positions and orientations of the end-effector tool [126]. The choice of joint configuration is crucial since the kinematic optimization of the manipulator is challenging owing to the inherent heterogeneity across tree designs and the available workspace (see Table 1.4).

Zahid et al. [127] created a manipulator for trimming apple trees with 2R and 3P degrees of freedom. The system can only support a single pose at each location in the workspace, with the manipulator effectively achieving the end-effector cutter's broad orientations. Zahid et al. [116] enhanced the manipulator by adding 3R and 3P DoFs for trimming apple trees after analyzing the limitation (Fig. 1.9p4). The combined end-effector was positioned using prismatic connections in both manipulator systems, and the revolving joints were attached directly to the end-effector to decrease the amount of space needed for manipulation.

The positioning joints often have a significant role in how a posture changes during manipulation, with the orientation (wrist) joints playing less. A joint arrangement that allows the manipulator a minimal posture change when navigating should be chosen since a greater degree of pose variation increases the risk of collision with branches. To lessen the chance of a collision with the branches, the manipulator design should consider aspects of the tree, such as canopy sizes, structures, and branch complexity.

(b) End-effectors

A crucial part of the pruning robot, the end-effector, is necessary to make the pruning cut on the chosen branches. Apple tree pruning end-effector tool design is complicated since the branches are often packed and overlapping, leaving little room for manipulation. According to Kondo and Ting [128], the design should consider the job object requirements, which include physical, horticultural, and biological features, as well as the mechanical and spatial requirements, which include size, shape, weight, and mobility. Branch and stem-cutting end-effectors with various architectures, joint configurations, and cutter processes have been created by several researchers [127, 129, 130]. The complexity of control required to cut depends on whether the cutter mechanism uses electric, hydraulic, or pneumatic power. Table 1.4

Table 1.4 Effectiveness of variously configured manipulators. Additionally, recently created end-effectors for cutting branches and stems

Reference	Actuation	Configuration (DoF, R/P)	Performance
[127]	Electric motors (custom)	5 DoF, 3P and 2R	High reachability, fewer spatial requirements
[116]	Electric motors (custom)	6 DoF, 3P and 3R	Reduced singularities, fewer spatial requirements
Reference	**Limitation**	**Cutting Mechanism**	**Performance**
[127]	Unable to cut the branches > 14 mm diameter	Pneumatic shear cutter	Cut the branches up to 12 mm diameter, 1 s
[116]	Void in the reachable workspace	Electric shear cutter	Cut the branches up to 25 mm diameter, 0.75 s per cut

P stands for the prismatic degree of freedom, whereas R is for the radical degree of freedom presents an overview of the branch and stem-cutting end-effectors related to apple fruits.

Zahid et al. [127] invented a pneumatic shear cutter end-effector to trim tall spindle 'Fuji' apple trees. Two rotating degrees of freedom were added to the end-effector to reduce the manipulator's space needs. The branch age and necessary cutting torque were found to have a 2×2 logical connection with an R2 of 0.93. According to the kinematic dexterity study, the end-effector cutter may be positioned at various orientation ranges, which is necessary for trimming branches. Branches up to 12 mm in diameter might be chopped using the pneumatic shear cutter. An electric shear cutter end-effector was created and tested in 'Fuji' apple trees trained to a fruiting wall design [116].

A branch-cutting dynamic study should be done to build the end-effector with the necessary cutting capabilities since the cutting torque requirements are for different apple kinds. The design should also consider the crop's other horticultural needs to ensure that the end-effector can create a clean cut. A rough cut might lead to branch deterioration from dry rot or interfere with healing. Another crucial design factor is the end-effector's adaptability for integrating various components like a camera, manipulator, as well as sensors.

(c) Automated Apple Pruning Techniques

Pruning tactics for fruit trees are primarily governed by guidelines that control canopy size and shape to enhance light distribution with the main objective of enhancing fruit quality. Removing a certain number of flower buds to regulate crop density is another purpose of pruning [131].

Alexander You et al. [114] presented a robotic system (Fig. 1.9p3) consisting of an industrial manipulator, an eye-in-hand RGB-D camera setup, and a specially designed pneumatic cutter. The system can organize a series of cuts while making a few environmental assumptions. They use a brand-new planning framework created for high-throughput operations that draw on prior work to shorten motion planning time and sequence cut points intelligently. With a high success rate for plan execution and a 1.5 x increase in throughput speed compared to a baseline planner, the system made a substantial advancement toward realizing robotic fruit pruning in practice.

The potential of pruning and training for managing various pests and diseases is also covered. In [112], they created and verified a 3D model of an apple tree during its dormant season (Fig. 1.9p1). Next, an algorithm was devised to identify pruning branches and the technique of identifying pruning branches was validated using accurate human pruning branch recognition. Branch identification and pruning branch identification algorithms have 76% and 86% accuracy rates, respectively. M. H. Korayem et al. [115] offer a fresh approach to the issue of automated and systematic dynamic modeling of wheeled mobile manipulators with dual arms. The equations of motion are constructed using the recursive Gibbs–Appell formulation to avoid calculating the "Lagrange multipliers" connected to the nonholonomic restrictions. Also, the 3×3 and 3×1 matrices, which are more effective than the 4×4 and 4×1, carry out all mathematical operations. They apply it to a mobile manipulator with two robotic arms to show how their approach derives the equations of motion for such complicated systems (Fig. 1.9p2).

For pruning apple trees, a two-degrees-of-freedom end-effector (2R) (Fig. 1.9p5) was constructed [117]. Then Branch diameter and cutting force for Fuji apple trees were shown to have a logical 2 × 2 connection ($R2 = 0.93$). Following that simulation, it was demonstrated that the cutter could be oriented in a spherical workspace in various ways. Finally, the created end-effector could cut branches with a diameter of up to 12 mm.

They presented an exact robotic pruning system (Fig. 1.9p6) that employs 2D RGB pictures and forces data to perform precise cuts with the least effort [118]. They show an increase of over 30% accuracy over a baseline controller using camera depth data using this simple but unique technique. In [119], the branch-cutting torque (Fig. 1.9p7) was assessed in this research because it will be crucial for developing a robotic pruning end-effector; a force-measuring sensor was coupled with a manual shear pruner to monitor the branch-cutting torque. And an inertial measurement unit sensor was also used to keep track of the angle between the shear blades and the branch. The maximum cutting torque of 6.98 Nm was found for 'Fuji' branches with a diameter of 20 mm for a straight cut with the branch positioned at the shear cutter center when all test results (four cultivars and cutting settings) were compared.

(d) Various analysis tasks of the apple pruning techniques

The many techniques for pruning and training deciduous fruit trees are described in [132], along with their effects on fruit output and quality, blooming, and cropping. The potential of pruning and training for managing various pests and diseases is also covered.

In the somewhat less performing YOLO-v5 model, the pruning inference may increase the inference speed to 71 FPS. In [133], branching availability was simulated for apple tree robotic pruning. An obstacle and kinematic manipulator model created a virtual tree environment. A rapidly exploring random tree (RRT) was integrated with smoothing and optimization for better route planning. The end-approach effector's angle and cutter orientation at the target were evaluated for their effects on RRT route planning. The simulations demonstrated that the RRT algorithm efficiently avoided obstacles and enabled the manipulator to reach the average target spot in 23 s. Smoothing and optimization helped cut the RRT route length by roughly 28% and 25%, respectively. The shortest route lengths were produced via the RRT smoothing technique; however, it added 1–3 s to the calculation time. In [134], they develop a pruning severity index derived by dividing the cross-sectional area of a tree's central leader at 30 cm from the graft union by the total of the cross-sectional areas of all branches at 2.5 cm from their cooperation to the significant leader. The main components of the 'Buckeye Gala' were gradually removed to change the limb-to-trunk ratio (LTR), which now has six severity levels ranging from LTR 0.5 to LTR 1.75. Lower values indicate more severe pruning with less whole-tree limb area than the trunk area. By controlling the canopy-bearing surface, the orchard manager may control crop load potential by using the LTR to determine the amount of pruning severity. Also, this measure is a crucial stage in creating autonomous pruning systems.

Xin Zhang et al. [135] examined precision pruning techniques that affected FRE in two sets of randomly chosen horizontal branches of 'Jazz/M.9' apple trees in a commercial orchard (106 and 107, respectively). FRE was substantially higher (91%) for branches treated with G1 than for branches treated with G2 (81%), and observed a negative correlation between FRE and the lateral shoot length. FRE ranged from up to 98% for shots less than 5 cm to just 56% for those 25 cm or longer. They created the shoot diameter-to-length index (S-index) to comprehend how to shoot size affects FRE. When the S-index was higher than 0.15, FRE reached a maximum of 98%. These findings imply that to attain an FRE of 85% without compromising fruit quality, lateral fruiting branches must be pruned to a length of less than 15 cm or an S-index larger than 0.03. [136] Aimed to determine differences in crop output, quality, and light dispersion between pruning techniques. The information was used to create a set of improved pruning 'rules' that a robotic pruner could easily adhere to. On the cultivars 'Pink Lady' and 'Golden Delicious,' they experimented with three distinct pruning techniques in two commercial orchards in Indiana. Commercial, statutory, and Purdue horticulture pruning were the three types of pruning (PHP). In [137], two tree pruning techniques were developed: tree pruning and major branch trimming. The parameters of tall-spindle fruit trees were extracted using contemporary laser scanning technology. According to the pruning findings, the DBH, tree height, and crown width relative errors were 9.27%, 4.35%, and 7.44%, respectively, and the RMSE was 8.78 mm, 143 mm, and 116 mm. Besides, crown width, tree height, space, and branch length were used to quantify the fruit tree's level of pruning depending on the two pruning techniques.

They employed the BP based on a neural network for apple fruit tree pruning and tested it using 40 sets of data from each tree [138]. Four hundred sets of data were used for model training and evaluation. The trial outcomes showed that the method's F1 score for the rear branches was 0913, its F1 score for the centripetal branches was 0.867, and its F1 score for the competition branches for the enhanced algorithm was 0.755. This simulation research [139] aimed to determine if a robotic manipulator with six rotational (6R) degrees of freedom (DoF) and a shear blade end-effector was branch accessible. In MATLAB, a simulation-focused tree canopy environment was created. The goal pruning points were reached by making a collision-free route using the Rapidly-exploring Random Tree (RRT) obstacle avoidance method. Path smoothing and optimization methods were also used to shorten the journey and determine the optimal way. The research offers the fundamental data for the following work on creating a robotic pruning system. T. L. Robinson et al. [140] contrasted three pruning techniques; the most drastic procedure included removing one to three branches and trimming back all of the tree's remaining fruiting branches or shoots by a third (stubbing back). A second treatment included removing 1–3 branches and removing 1/3 of all fruiting spurs (spur pruning or extinction). A third procedure included removing 1–3 branches without further trimming (minimal pruning). The stubbing back treatment and spur extinction pruning significantly decreased floral bud load, fruit quantity, and yield compared to minimum pruning. Nevertheless, they boosted fruit size and crop value.

Azlan Zahid et al. [141] examined branch-cutting torque and angle while considering how a robotic end-effector would grow. The developed method included a manual shear pruner with a force-measuring sensor that could track the forces the user's hand applied. An Inertial Measurement Unit (IMU) was further employed to capture the shear blade's orientation. The findings also revealed that the cutting angles for Fuji apple trees did not significantly vary in the amount of cutting torque needed. In contrast, the tension required to cut branches from Fuji apple trees is greatly influenced by the point at which the branch and blade make contact.

The DH-parameter technique created the kinematic model for the manipulator to prune apple trees. MATLAB models for the manipulator and the virtual tree canopy were made [142]. An end-effector-equipped six-rotational (6R) degrees of freedom (DoF) robotic manipulator's workspace and collision-free route planning were investigated. The two-way rapid exploration random tree (RRT-connect) technique created a collision-free route to the desired pruning point. This study establishes a specific foundation for robot path planning used in apple tree trimming.

The growth and production are crucial aspects of agriculture and the significance of general pruning on apple fruits. Regarding our study on the mechanical robotic pruning of apple fruits from 2010 to 2023, we see that the advancement was quick, and the results were great. However, there is still room for improvement in the hardware and the software, the capacity to handle very large and varied branches, and speed and execution time (see next section).

1.3.3 Obstacles and Future Perspectives of the Apple Pruning Robotic

Technology for pruning is extensively employed in forestry and agriculture. Pruning may enhance the environment's beauty and encourage the development of plants. Agriculture and complex forestry pruning still rely on labor-intensive human labor, and the development of pruning robots is progressing slowly. Mechanization of forestry and agriculture has recently gained popularity worldwide, considerably aiding the development of pruning robots.

To manage tree growth, fruit quality, and production, branches are pruned. Adopting the right amount of pruning also helps with pest and disease management. Apple trees are traditionally pruned by hand; however, the procedure is costly and labor-intensive, and the pruning choice differs from person to person depending on one's talents and experience. However, these professionals' availability has diminished during the last several years. The agriculture sector is seriously concerned about the workforce shortage in the business. Several studies have examined mechanical pruning, or hedges on fruit trees, as a solution. After that change, mechanical pruning functions effectively, although it faces several difficulties due to the intricate structure of apple trees, speed, and variations in size. Route planning, pruning cut sequence, manipulation control, and obstacle avoidance are also crucial for the automated

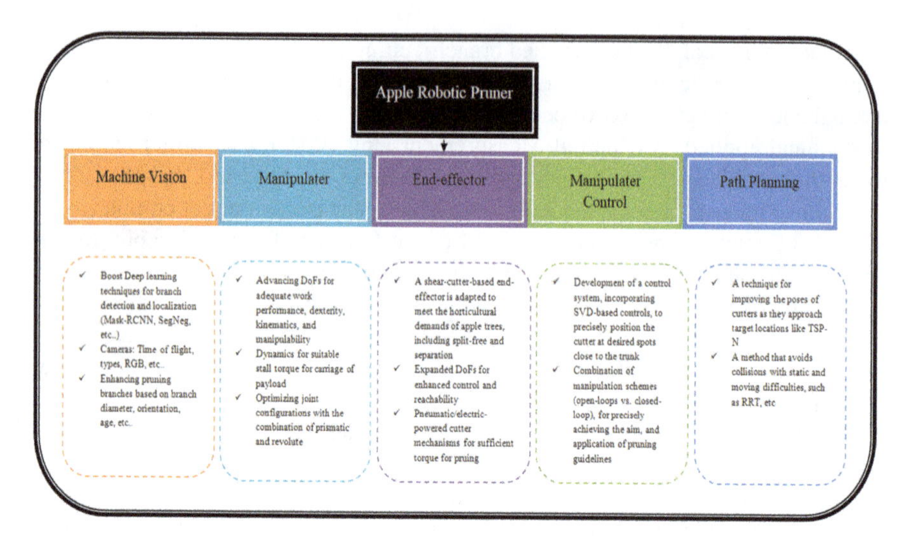

Fig. 1.10 A visual summary and future views for the apple robotic pruners development

pruning system to function properly. Figure 1.10 shows our perspectives and future reconditions in various stages of automated applied development.

1.4 Apple Fruits Thinning

The fruit thinning removes some of the harvests before it ripens on the tree to boost its continuing viability for sale and break its biannual bearing pattern [143]. Fruit trees generate more blossoms than are needed for the commercial harvest, which results in higher yields, better-quality fruits, and larger fruits. Floral thinning controls the amount of fruit that will be set, consequently impacting the fruit's quality and size. Dilution is also necessary to compensate for certain fruit crop kinds' propensity to yield fruit every other year. At the moment, crop thinning is done manually or using pesticides. Extension specialists and crop consultants employ flowering numbers when scheduling dilution sprays as a component of crop load management after chemical dilution. To forecast lightness in apples, two models are utilized. Inflorescences are required for nebulizer inputs [144]. Apples are defined by their heavy fruit production and intense blooming throughout the growing season, which has several unfavorable effects related to their tiny size, subpar color, and poor quality fruits. Additionally, the next year's flower bud growth was dramatically reduced, which led to lower yields of inferior fruits with a shorter rate of after-harvest storage life [145]. Apples may be thinned in various ways: by hand, chemically, and mechanically [146].

Thinning and floral/fruit development have been the subject of many review publications exploring mitigation impacts [147–153] it covers the many forms of thinning, their applications, and their historical development. Since the apple fruit has such high value and economic significance, there is a wealth of information about apple thinning; our attention will be on the difficulties thinning procedures face and the creation of cutting-edge, intelligent equipment to improve quality and boost production.

1.4.1 The Development of the Automated Apple Thinning

Fruit thinning is a management technique that decreases the number of fruits per tree in the current season, increasing the size of the fruit that remains on the tree and resulting in greater blooming and yield in the next season. Apple thinning is the most crucial management tactic in figuring out how profitable an apple orchard will be each year [154]. Therefore, to enhance efficiency and schedule real output, it is essential to know the difficulties associated with apple thinning and the supporting elements and smart mechanisms.

In [155], two mechanical thinning tools (Fig. 1.11t1) applied to the various types of fruits showed excellent results and improved the quality of apple production. The new technique (Fig. 1.11t2), which consists of three horizontal rotors, thins apple blooms precisely at various locations inside or outside the upper or lower tree canopy to increase fruit quality by eliminating fruit that would otherwise be shaded or excessive [156]. The technology reduces labor costs that would otherwise be associated with hand thinning by about 15 to 30 h/ha. The tool could also work with other fruit crops where fruit thinning is necessary to create high-quality fruits. In [157], three movable horizontal rotors make up the mechanical thinning apparatus, and their vertically spinning brushes use centrifugal force to remove a variable number of blooms. The front three-point hitch of a tractor, autonomous vehicle, or OTR is where the horizontal rotors are mounted to a vertical axis supported by a platform and evaluated on two different fruit trees: a Golden Delicious Reinders tree that is 18 years old and a Gala Mon-dial tree that is eight years old. The fruit was 17% larger, firmer (8.4 kg cm^2 in Gala vs. 7.6 kg cm^2 in the unthinned control), sweeter (124 vs. 117 g kg^1 sugar in control), had the highest malic acid content (4 g kg^1 vs. 3.4 g kg^1 in control), and contained 17% more anthocyanin (normalized anthocyanin index = 0.8 in Gala vs. 0.7 in control). Dennis Hehnen et al. [158] developed new mechanical thinning equipment (Fig. 1.11t3) and evaluated it on 288 ‘Buckeye Gala’ apple trees that were seven years old at a 1.2 m 3.7 m distance. The conventional chemical standard (105%) produced the best return bloom, which was followed by the strongest mechanical thinning (92% with 360 rpm), and then the combination of mechanical and chemical thinning (85%), which was superior to the control’s much lower values (69%). This demonstrates the effectiveness of blossom thinning in overcoming alternate bearing. In [159], the outcomes of this two-year investigation show that MBT may significantly lower apple trees’ fruit set. As MBT’s overall impact on yield characteristics varies annually, it should be used in addition to current chemical thinning techniques. The production and fruit quality was unaffected by

Fig. 1.11 Automated apple thinning equipment: (t1) Mechanical apple thinning tools [155]; (t2) The new thinning device's schematic [156]; (t3) Newly developed apple flower thinner [158]; (t4) Stem-cutting end-effector prototype [160]; (t5) An automated platform was utilized to scan the apple trees branch by branch [161]; (t6) Apple fruitlet thinning robot MARK 2

damage to the spur leaves, but further study is needed to fully assess the dangers that MBT may pose to the spur leaves.

The combination of NAA at a low rate with mechanical thinning strongly reduced fruit sets in 'Gala' and 'Jonagored' and produced a bigger increase in fruit weight and yield in 'Pinova' than hand thinning plus mechanical thinning [162]. Based on these findings, they concluded that mechanical thinning is a viable choice for apples. The results varied based on the device setting and cultivar. In 'Gala Must' and 'Jonagored,' NAA outperformed mechanical thinning in efficiency. Jaume Lordan et al. [163] investigated several agents and mechanical thinning in this research to provide an alternative to traditional thinners. The findings of the overall investigation indicated that olive oil might thin fruit, but its rate must be controlled to prevent fruit rusting. Hand thinning may attain appropriate levels; mechanical thinning provides more consistent outcomes than chemical thinning. Uddhav Bhattarai et al. [164] utilized Mask R-CNN, based on the Feature Pyramid Network (FPN) with ResNet101, on a collection of 205 images with a 1920 × 1080 pixels resolution. The detection's highest average precision (AP) is up to (0.86%) integrated with the machine vision system to assist automated apple thinning performance. A stereo-vision system was created [165] to size fruits on a tree and measure their seasonal development by matching fruits in photographs taken at various times. Fruit identification and on-tree fruit size were accomplished using neural network models such as Faster R-CNN and Mask R-CNN. The top fruits with an average diameter of 25 mm also showed a 73% matching accuracy, helping to operate thinned apples. In [160], their main objectives were to examine the dynamics of green fruit removal and to create and test a prototype end-effector for thinning green fruit. First, they ascertained the force requirements for pulling and stem-cutting green fruit removal; then they designed and developed a stem-cutting end-effector prototype; and finally, they tested the developed end-effector in an orchard environment by integrating the end-effector to two different base mechanisms, namely, a handheld bar and a robotic manipulator (Fig. 1.11t4). All end-effector prototype studies with green fruit removal had success rates above 90%.

Fruit and stem segmentation and orientation estimation methods were developed and tested for suitability in a machine vision system for robotic green fruit thinning in apple orchards [167]. The average accuracy for detecting green fruit and stems was 91.3% and 67.7% for mask sizes bigger than 322, respectively, demonstrating the Mask R-CNN model's success in segmenting green fruit. The stem orientation estimate attained extremely high accuracy with equivalent ratings of 99.8% and 99.7%. The vision system for an autonomous apple fruitlet thinning robot is presented in [161], along with its basic design, implementation, and assessment details. The platform comprises a robotic UR5 arm and stereo cameras that allow it to see through the leaves and map the exact quantity and size of the fruits on the apple branches (Fig. 1.11t5). They demonstrate that this platform can estimate the apple tree's fruitlet load with 84% accuracy and 87% precision in a real-world commercial apple orchard.

1.4.2 Challenges and Opportunities

Fruit thinning is one of the crucial steps in boosting the production, development, and blooming of high-quality apple fruits. Apple fruit thinning encourages early fruit dropping and dictatorship; it enhances the fruit's color, size, and remaining quality. It helps prevent limb harm from heavy fruit-bearing; it boosts the harvest the next season.

In the present research, we focus on reviewing the thinning of apple fruits; the story begins with manual thinning and the contribution of the first to improving apple fruits and production. Still, it results in more labor challenges and slowness. It is impractical for large areas of orchards, followed by chemical and mechanical thinning, which increases work efficiency and adds more improvements compared to manual thinning, and then the emergence of mitigation by intelligent robots with more speed, accuracy, and quality in the various thinning processes.

Several studies in this section on different levels and types of apple fruits in mechanical [155–162] and automated chemical dilution [163] and at the level of robots [164–161] achieved excellent results, which contributed to increasing production and quality. We observed the models that applied to the vision system: Mask R-CNN was based on a Feature Pyramid Network (FPN) with ResNet 101; Mask R-CNN plus Principal Component Analysis (PCA); and Mask R-CNN uses ResNeXT-101 for image training with the RANSAC algorithm for the evaluation. The most used model was Mask R-CNN, and the model with the highest performance was Mask R-CNN + PCA, with an accuracy of (99.8%). The highest robot performance was [167], with success rates over 90%.

The crucial yet challenging technique of fruit thinning mostly drives profitability in orchards. The primary technique growers utilize is chemical thinning because of high labor expenses and the inability of hand thinning to increase return bloom. Mechanical thinning is a good alternative to chemical thinning, although it is unpredictable and has safety and environmental problems. After that, robotic thinning comes with cost challenges and maintenance.

There are several difficulties in implementing robots and automation solutions for this orchard work because of the tiny size of the target, the blossoms, compared to the enormous dimensions of the parachute. Robotic instrumental thinning has been centered on floral thinning, with the introduction of components of mass removal systems' independence, such as string relaxers and robots' development. Arm and end effectors thinned. Additionally, efforts were made to identify flowers in color or multispectral photos for two purposes: precise estimations of blooms as inputs to dilution models and 3D estimation of canopy and flowers for automated reduction [137].

There are many opportunities for development at the mechanical and intelligence levels of machines; the below represents our observation based on the existing research: (a) Vision systems should be added to the mechanical devices to speed up performance and improve quality, and the circumference of the cutting tools should be increased to be able to cover a large number of branches. Additionally, test several apple blossoms at various ages to further the assessment study and improve the equipment's speed by adding more rotors; (b) decreasing undercounting through sphere fitting method advancements and system upgrades is still necessary. It is still necessary to compare the size estimations to ground truth measurements. The fruitlet counting approach and a size estimate and quality evaluation of fruits may help to obtain successful fruitlet thinning outcomes; (c) the ability of the green robotic fruit system to extract fruit in various positions and orientations is increased by integrating with several robotic manipulators. A computerized vision system will probably be developed and coupled with the end actuator to create an automated green fruit-slicing framework. Besides, as an outcome of the poor matching efficiency, performance could be improved by modifying new functions with quick processes to speed up the process and application and enhance the detection quality. Finally, adding control devices and sensors can upgrade the mechanical thinning machine's performance.

We reach the following conclusion: the primary previously mentioned points offer means for keeping pace with the acceleration of development according to the existing capabilities, which helps to improve output and the quality of apple fruits.

1.5 Apple Fruits Pollination

Most apple trees need pollination to produce fruit; hence, it is crucial. Apple trees must exchange their chromosomes, just as people do. Although certain self-fertile apple trees may bear fruit without the assistance of a pollinator, every tree benefits from having a companion [165]. Apple producers must plant more flowers around their orchards to boost their production. The transfer of pollen from one tree to another is necessary for pollination. Typically, bees 'and other pollinating insects' are responsible for this. (When things are unstable, you may do this with a paintbrush [167]).

One of the essentials of producing apples profitably is pollination. Since most apple types are not self-fertile, they need a suitable pollinator to make a complete harvest. A fruit's ultimate size and quality are influenced by successful pollination, which produces numerous healthy seeds and higher yields than when pollination is unsuccessful. Selecting a pollen type with suitable pollen grains and an overlapping blooming season is crucial [166]. Regarding the pollination procedures and advantages of fruits and crops, we discover that manual and automated pollination occurs. Control over pollen source, amount, time, frequency, and independence from environmental changes are advantages of hand pollination. It seeks to enhance fruit quality, prevent fruit abortion, create employment possibilities, and provide access to food for subsistence. The primary barriers to hand pollination are high labor expenses, high material prices, and the need for specialized knowledge [168]. In March 2018, Wal-Mart submitted a patent for an autonomous robotic bee, demonstrating the growing popularity of intelligent pollination as a solution to the pollinator decrease issue. However, she clarifies that this 'solution' is technically and economically unworkable at the time [169]. The developments of clever procedures to successfully address the transient issues of pollination, speed, and quality to boost future output were then seen in research sources and institutions.

1.5.1 Overview of Automated Apple Fruits Pollination

Most fruit and nut crops need pollination to produce, although it is often a limiting factor for productivity and product quality. A variety of horticultural fruit and nut crops have the potential to become more productive thanks to mechanical pollination (MP) systems, which reduce the risk of reliance on the currently available insect pollination operations [170]. The development has been the subject of research in various fruits [76–83], highlighting the need to create autonomous and intelligent pollination techniques. In this section, we specifically concentrate on the development of apple fruits.

Xinyang Mu et al. [172] initially created a dataset on apple blossom blooming stages from first king to full bloom. A deep learning system was then designed to detect and identify the king flower, allowing researchers to estimate how much it blossoms and where it is in the tree canopy. The study's findings are anticipated to offer robotic pollination decision-making data, horticulture expertise, and the present climatic condition. In [171], they suggest establishing an automated apple crop load management system to produce fruit harvests with high yield and superior quality, which would significantly increase the net profit of the fruit business. The proposed automated system consists of two main parts: a sophisticated robot vision system that can locate clusters of apple flowers and monarch flowers, map flower densities, and automatically interact with the robotic handler, and a cartesian robot system that can move to the desired location. An automated crop load management system for the apple product is anticipated to be created using the prototype and field test as a guide.

Stress, torque, and safety factor evaluations have been used to improve the design and operation of a drone system to ensure that the final product can survive the stresses imposed during the process [173]. Several drone kinds and plant pollens will be able to be used with this drone pollinator. The application boosted fruit quality and pollination system efficiency.

To identify and find the king flowers from an apple flower dataset throughout the blooming stage, from initial king bloom to full bloom, a Mask R-CNN-based detection model and king flower segmentation technique were built [174]. Regarding flower phases of 20% to 80% blooming, the king flower recognition accuracy ranged from 98.7% to 65.6%. The study's findings are anticipated to provide robotic pollination judgment data and horticulture knowledge.

We now reach the section on deep learning models to improve the visual system for the tools used with apple software. Utilizing models based on deep learning with exceptional representational capabilities, such as CNN and SVM techniques, both studies offered novel approaches for identifying apple blossoms [175, 176]. To categorize the growing stage of apple blossom, YOLOv4, an advanced CNN-based object detector, has been released [177]. For semantic picture segmentation and active figure perimeter implemented by the level ensemble, another method for flower recognition using a collection of deep learning algorithms has been given [178]. A promising approach for blossom identification in recent years has been Mask R-CNN. Apple flower detection using the deep learning method based on Mask R-CNN was observed. A better Mask R-CNN with a U-Net backbone 'referred to in the paper as MASU R-CNN' is suggested using the robust segmentation of instances framework to segment apple blossoms into pink bud, semi-open, and completely open phases inside an environment [179].

In contrast to the other automated processes of the apple fruit, we note that there is a need for more development and better quality hardware, robotics, and gadgets linked to apple pollination. In the software components and applications, we observe that conventional techniques like SVM, RF, NB, and CNN for deep learning are still used in the vision system for the current applications, Mask R-CNN. Therefore, there are great opportunities to develop the tools needed for apple pollination, including the drone device and bees [169], with advanced improvements to the types, forms, speeds, determinations, localization, and other control materials. There need to be more software models and resources; first, it is possible to add more datasets related to apple blossoms; second, customize new platforms that can handle substantial corpora with high efficiency—applying the abovementioned key roles for developing and upgrading apple pollination machines to improve performance, production, and quality.

1.6 Apple Fruits Bagging

Bagging is a physical protection that may change the fruit's microenvironment and boost the fruit's appearance by lightening the skin tone and minimizing imperfections. It may enhance the fruit's interior quality in many ways [180]. In addition to lowering the risk of illnesses, insect pests, mechanical damage, sunburn on the skin, fruit breaking, pesticide residues on the fruit, and bird damage, bagging fruit before harvest may also lower the risk of skin burn [181].

Multiple factors affect the bagging, such as physiological factors influenced by apple fruit, biotic factors influenced by apple bagging, and physiological and biochemical factors, as shown in (Fig. 1.12), including the sub-variable for each aspect, that assist the quality of the apple bagging. Apple fruit bagging is a costly, time-consuming, and laborious procedure that comprises several steps. These constraints have somewhat constrained the creation and implementation of fruit-packing technology [182].

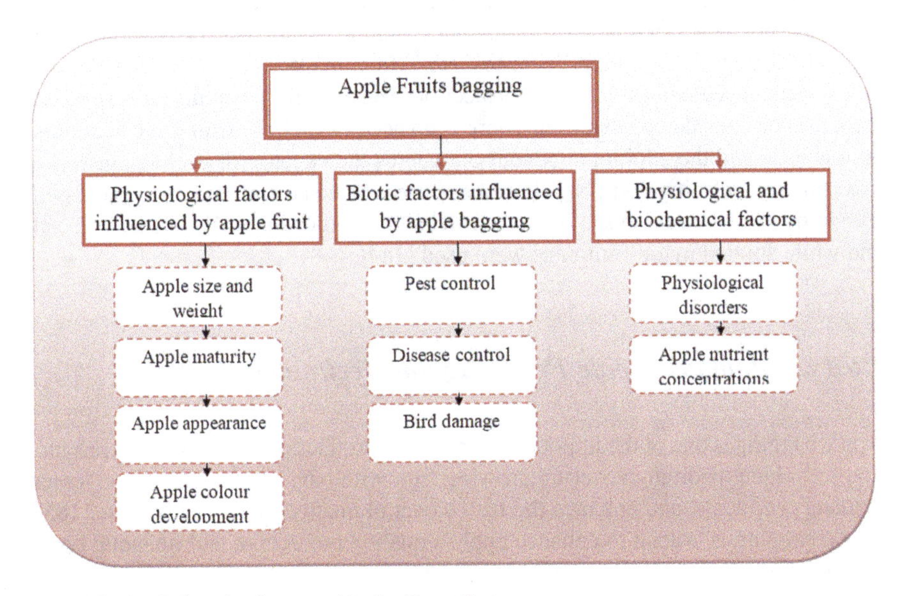

Fig. 1.12 Apple bagging impact with significant factors

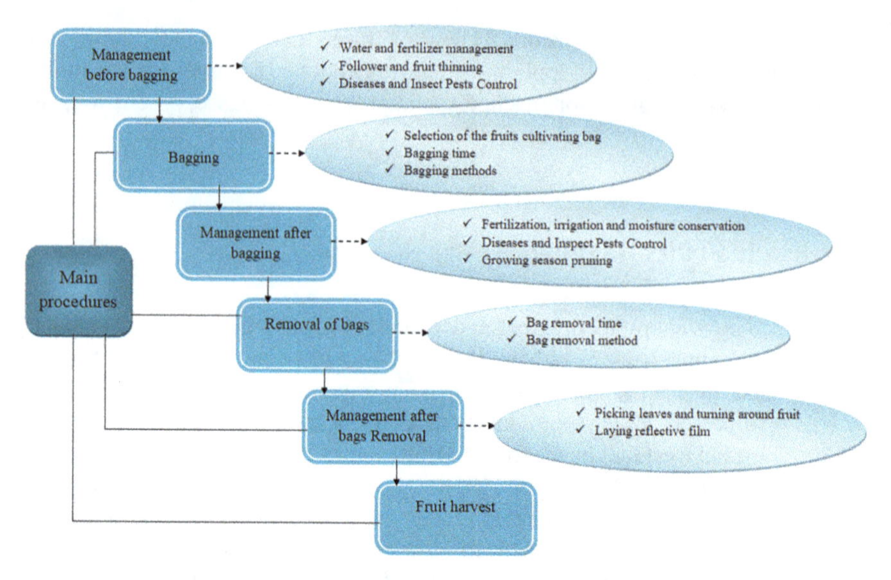

Fig. 1.13 Primary apple bagging procedures

Fruit packaging generally consists of management before bagging, bagging, and controlling after bag removal [183]. The major relationships and bagging procedures are briefly outlined in (Fig. 1.13), which uses apple bagging as an instance.

Fruit bags must often be used in greater quantities to protect fruits with thinner skin, like apples, pears, grapes, peaches, etc. Often, thick-skinned fruits like durian and grapefruit need simple pesticide protection. The financial advantages of fruit trees are a further crucial consideration. There is no need to pack the fruit if the economic gain is little and the additional benefit outweighs the expense of packaging. Nylon and non-woven fruit growth bags have been used due to advancements in packaging technology and the demands of various fruit bagging procedures. Among them, black and white applied nylon fruit bags were used [184].

1.6.1 Automated Apple Fruit Bagging Technologies

Apple bagging is one of the important steps in the production of fruits. It is a production technique involving covering growing fruit with bags to keep out pests, lessen residual chemicals, and enhance the fruit's overall quality and attractiveness [185]. Apple bagging is crucial to enhance apple's quality, protection, and financial benefits to the economy. Consequently, the materials and current technology will make it essential to continue developing, making the operations shorter turnaround times, cheaper costs, and more efficient [186].

In [187], they investigated a vision-based localization apple bagging robot for young apples (Fig. 1.14b1). In the visible light of the natural environment, the fundamental technologies for stereoscopic picture placement and recognition of young fruit are investigated. The gathered photos of young apples are initially preprocessed using the Otsu segmentation technique. Second, the preprocessed pictures of young apples are marked and labeled using the enhanced connected component labeling method. Finally, a binocular stereo-vision system positions the apple fruit in three dimensions. Wang et al. [188] employed the Arduino MCU as the control center for the new fruit tree bagging equipment (Fig. 1.14b2). They have enhanced bagging effectiveness while lowering farmer labor input and agricultural product labor input. In addition, the innovative bagging machine has a handle and a telescopic rod conversion tool that can be adjusted to perform multilayer bagging. Due to the tailored fruit, the innovative bagging machine employs widely available fruit bags, lowering farmers' production efficiency and costs.

Liu et al. [194] used a watershed technique to split the source pictures into asymmetrical blocks based on the findings of edge detection on R-G grayscale images. The segmentation approach may retain fruit edges and minimize the number of blocks by 20.31% compared to the watershed algorithm based on gradient pictures. The support vector machine then divides these blocks into fruit and non-fruit blocks based on the color and texture information retrieved from the unions. The outcomes of the suggested method's FNR and FPR are 4.65 and 3.50%, respectively. Gou et al. [189] introduced the four-rotor flying system's general design architecture (Fig. 1.14b3). Next, the ground coordinate system and the collective coordinate system are developed for the dynamic model of the four-axis aircraft. Obtaining reaction time, control stability, and robustness all occur simultaneously. The design of an aircraft's control algorithm uses sliding mode variable structure control theory. Lastly, it is shown that the controller's strength and stability satisfy the project's criteria by comparing it to the controller using simulation data. In [190], a unique providing mechanism (Fig. 1.14b4) was suggested based on the manually operated fruit bag case. The providing device's open hand works similarly to a farmer's hand, which can continually pull out multiple-layer fruit paper bags one at a time and entirely open them from the inside. In the investigation, the operating capability of the prototype device was examined at various driving trajectories and speeds. The created providing device could accomplish more than 90% opening success rate with the synchronous driving trajectory and permitted high-moving speed.

A mechanical design for sorting and packing apples is put forth [191]. Each component's purpose is thoroughly explained, as well as the machine's overall structure (Fig. 1.14b5). Functional analysis is performed on the bagging, intelligent classification, wax drying, and poor fruit removal modules. Less surface damage to the apples and more automation are the goals of the design strategy. Gou et al. [195] described a specific plastic box bag that shielded apple fruits from hail. Apple bags are finished in 1 s, and the pace of the bags rises by 12 fold. This kind of apple bag was discovered via extensive study and testing to be inexpensive, quick, and reusable. It may also reduce fruit rust and prevent illnesses, insects, birds, rodents,

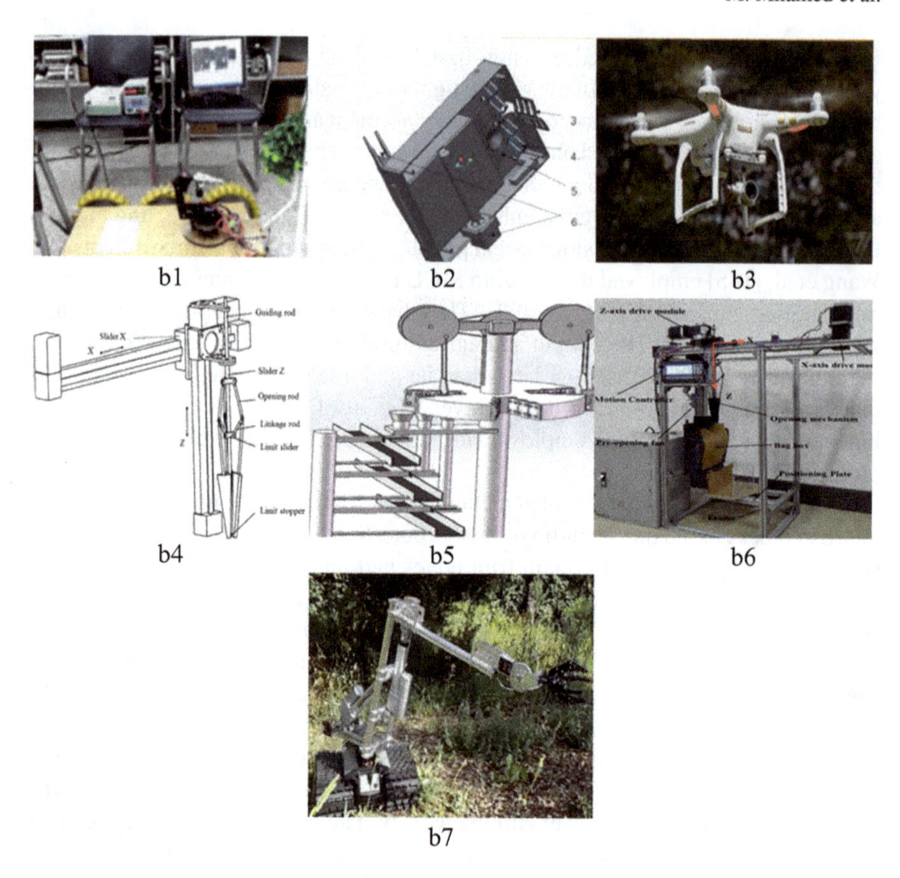

Fig. 1.14 Automatic apple bagging machinery: (b1) Apple bagging robot platform [187]; (b2) Apple bagging device [188]; (b3) Wholly regulated by the four motors' speed [189]; (b4) Prototype of the bagging device [190]; (b5) apple's packaging and sorting machinery [191]; (b6) Examine platform [192]; (b7) Anti hail apple bagging robot [193]

bees, and other pests from hurting the fruit. It also reduces friction between leaves and branches, protects against sunburn, and soothes hail and other mechanical wounds.

Mengsheng Zhang et al. [196] explored the viability of utilizing the starch index as the maturity index in visible and near-infrared (Vis-NIR) spectroscopy to assess the maturity of bagged 'Fuji' apples. Eight hundred forty-six apples were sorted into three ripeness categories using the starch index (immature, harvest mature, and eatable mature). The primary components of distinctive wavelengths in the spectral data were extracted using principal component analysis, the random frog (RF) method, and the RF algorithm coupled with the successive projection algorithm (RF-SPA). Comparing the outcomes of various modeling techniques, it was found that the RF-SPA-LSSVM model, based on 15 characteristic wavelengths, had the best prediction performance. The prediction set's classification accuracy was 89.05%, and the three varieties of apples' areas under the receiver operating characteristic curve exceeded 0.9210.

Reference [192] aimed to establish a rigid-flexible coupling model with the RecurDyn software to simulate the contact action of the open mechanism on a typical ordinary multilayer fruit paper bag and evaluated bag open effects by simulation and experiment (Fig. 1.14b6). They discuss the effectiveness of the simulation method for evaluating the contact action effect of the developed open mechanism on the multilayer fruit paper bag; after that, they investigate the main influence factors for the opening effect and achieve the law of how the available mechanism affects on stress distribution, overall deformation of the multilayer fruit paper bag. Gou et al. [197] primarily study the quality characteristics and viability analysis of Gansu's 5000 Mu hail-proof plastic bags. The effectiveness of hail-proof plastic bags was quick, increasing job effectiveness, lowering environmental pollution, and lightening the labor of fruit farmers. In [193], they solved and analyzed the forward and inverse kinematics of the anti-hail apple-bagging robot manipulator (Fig. 1.14b7). The robot toolbox based on MATLAB is used for simulation verification and joint space trajectory planning simulations. The kinematics model is established based on the standard matrix representation.

Yuan et al. [198] presented the design of an automated 8-cm shock-proof netting machine. The gadget fits apple fruit shockproof net sleeves automatically. It is based on the Siemens PLC-S7-200 SMART ST20 model, which includes a bagging device, net storage device, cutting device, photo-electric sensor acquisition, and electrical control. The bagging device fits the anti-collision net sleeve once the sensor determines that the apple has reached the predetermined location. The electric heating wire-cutting device subsequently trims off the extra anti-collision net sleeve. This machinery can efficiently save labor, lighten the load on an aging society, and lower the cost of manufacturing.

1.6.2 The Potential Future Development for the Automated Apple Bagging

Agriculture, in general, and the production of fruits and vegetables, in particular, depend heavily on bagging to dramatically improves the form, size, look, and quality of fruits like apples and others, resulting in quicker marketing, higher-quality products, and more output.

The journey starts with creating bags [195, 197] that follow the required standards, considering the natural elements important in various regions and the type of fruit for high quality. Next, the additional tools in the bagging operations and contemporary mechanisms and their effectiveness, speed, and periods are discussed, along with the investigation findings. The review was split into three categories: A group focused on the ingredients used to make the bags and how they affected the apple fruit [187, 188, 189, 191]. To improve physical hardware and boost speed and accuracy [191, 196], some organizations concentrate on intelligent mechanisms and their parts, development based on them, and performance [194, 190, 196, 192, 193, 198].

Based on the earlier research and we notice the following about prospective future developments for automated apple bagging: (i) extending the picture datasets for a more excellent demonstration of the proposed techniques and success rate effectiveness, as well as upgrading the methods to extract the best feature that aids in improving performance, are both conceivable. Additional investigation using several fruit varieties is required, besides the possibility of adding new functions and developing the vision system's applications; (ii) it's possible to give more concern to the localization and control devices. Its structural stability and driving reliability must improve its supply rates. Additional sensor and localization items are needed to enhance the automated apple bagging, which is integrated with an arm device that helps speed up the bagging process; (iii) more apple samples, such as different types of apples from various places, must be obtained to increase the model's universality. And need more comparison with the latest deep learning approaches to see the ability to extract the features; (iv) to produce the finest bag opening effects, the sliding stroke of the opening mechanism should be carefully managed. A faster sliding speed is desired within the mechanical capacity and supply device driving limits. It may provide a way to identify the open mechanism's operating characteristics and enhance the application performance of the supplying devices.

There are still difficulties, such as the high material costs of creating physical hardware, the application of artificial intelligence models in software development procedures, and the shortage of studies that allow us to work on the development, progress, and provision of intelligent economies that aid in accelerating and enhancing apple fruit production and economic benefits. All the gaps mentioned will provide more opportunities and rapid developments for researchers' communities.

1.7 Conclusion

This paper surveyed the automated development of apple fruits in five directions: picking, pruning, thinning, pollination, and bagging. Following that, we discussed the most beneficial tools and the most effective software and highlighted the hardware and software gaps and the key challenges. We summarized the robot components, performance throughout time, applications utilized, resources, and discoveries. After that, we listed significant developments, strategies, findings, and gaps. Next, we discussed the specific challenges. Lastly, we provided perspectives, opportunities, solutions, and future views. These all contribute to continuing and keeping up with the development of apple fruit production, mechanisms, processes, and quality.

References

1. Javaid M, Haleem A, Singh RP, Suman R (2021) Substantial capabilities of robotics in enhancing industry 4.0 implementation. Cogn Robot 1:58–75
2. Zhang Z, Lu R (2021) Automated infield sorting and handling of apples. Fundam Agric Field Robot:267–295
3. Vougioukas SG (2019) Agricultural robotics. Annu Rev Control Robot Auton Syst 2:365–392
4. Bechar A, Vigneault C (2017) Agricultural robots for field operations. part 2: operations and systems. Biosyst Eng 153:110–128
5. Bogue R (2016) Robots poised to revolutionise agriculture. Ind Robot Int J 43(5):450–456
6. Basu S, Omotubora A, Beeson M, Fox C (2020) Legal framework for small autonomous agricultural robots. Ai Soc 35:113–134
7. Zhang Z, Igathinathane C, Li J, Cen H, Lu Y, Flores P (2020) Technology progress in mechanical harvest of fresh market apples. Comput Electron Agric 175:105606
8. Zhang Z, Pothula AK, Lu R (2017) Development of a new bin filler for apple harvesting and infield sorting with a review of existing technologies. In: 2017 ASABE Annual International Meeting, American Society of Agricultural and Biological Engineers, p 1
9. Zhang Z, Heinemann PH, Liu J, Baugher TA, Schupp JR (2016) The development of mechanical apple harvesting technology: a review. Trans ASABE 59(5):1165–1180
10. Roldán JJ, del Cerro J, Garzón-Ramos D, Garcia-Aunon P, Garzón M, De León J, Barrientos A (2018) Robots in agriculture: state of art and practical experiences. Serv Robot:67–90
11. Gil G, Casagrande D, Cortés LP, Verschae R (2023) Why the low adoption of robotics in the farms? challenges for the establishment of commercial agricultural robots. Smart Agric Technol 3:100069
12. Zhu CW, Hill E, Biglarbegian M, Gadsden SA, Cline JA (2023) Smart agriculture: development of a skid-steer autonomous robot with advanced model predictive controllers. Robot Auton Syst:104364
13. Zhang Z, Pothula AK, Lu R (2017) Economic evaluation of apple harvest and in-field sorting technology. Trans ASABE 60(5):1537
14. Hanif MK, Khan SZ, Bibi M (2023) Applications of artificial intelligence in pest management. In: Artificial intelligence and smart agriculture applications. Auerbach Publications, pp 277–300
15. Ahmad MN, Anuar MI, Abd Aziz N, Bakri MAM, Hashim Z, Seman IA (2023) Addressing agricultural robotic (agribots) functionalities and automation in agriculture practices: what's next? Adv Agric Food Res J 4(1)
16. Pedersen SM, Fountas S, Have H, Blackmore B (2006) Agricultural robots—system analysis and economic feasibility. Precision Agric 7:295–308
17. Pedersen SM, Fountas S, Sørensen CG, Van Evert FK, Blackmore BS (2017) Robotic seeding: economic perspectives. Precis Agric Technol Econ Perspect:167–179
18. Blackmore B, Fountas S, Gemtos T, Griepentrog H (2008) A specification for an autonomous crop production mechanization system. In: International symposium on application of precision agriculture for fruits and vegetables, vol 824, pp 201–216
19. Fountas S, Gemtos T, Blackmore S (2010) Robotics and sustainability in soil engineering. Soil Eng:69–80
20. Marinoudi V, Sørensen CG, Pearson S, Bochtis D (2019) Robotics and labour in agriculture. a context consideration. Biosyst Eng 184:111–121
21. Ghobadpour A, Monsalve G, Cardenas A, Mousazadeh H (2022) Off-road electric vehicles and autonomous robots in agricultural sector: trends, challenges, and opportunities. Vehicles 4(3):843–864
22. Shamshiri RR, Weltzien C, Hameed IA, Yule IJ, Grift TE, Balasundram SK, Pitonakova L, Ahmad D, Chowdhary G (2018) Research and development in agricultural robotics: a perspective of digital farming
23. Zhai Z, Martínez JF, Beltran V, Martínez NL (2020) Decision support systems for agriculture 4.0: survey and challenges. Comput Electron Agric 170:105256

24. Santos LC, Santos FN, Pires ES, Valente A, Costa P, Magalha˜es S (2020) Path planning for ground robots in agriculture: a short review. In: 2020 IEEE International Conference on Autonomous Robot Systems and Competitions (ICARSC). IEEE, pp 61–66
25. Vasconez JP, Kantor GA, Cheein FAA (2019) Human–robot interaction in agriculture: a survey and current challenges. Biosys Eng 179:35–48
26. Mahmud MSA, Abidin MSZ, Emmanuel AA, Hasan HS (2020) Robotics and automation in agriculture: present and future applications. Appl Model Simul 4:130–140
27. Albiero D (2019) Agricultural robotics: a promising challenge, Curr Agric Res J 7(1)
28. Hajjaj SSH, Sahari KSM (2016) Review of agriculture robotics: practicality and feasibility. In: 2016 IEEE International Symposium on Robotics and Intelligent Sensors (IRIS), IEEE, pp 194–198
29. Xu R, Li C (2022) A review of high-throughput field phenotyping systems: focusing on ground robots. Plant Phenomics 2022
30. Wang F, Ge S, Lyu M, Liu J, Li M, Jiang Y, Xu X, Xing Y, Cao H, Zhu Z et al (2022) DMPP reduces nitrogen fertilizer application rate, improves fruit quality, and reduces environmental cost of intensive apple production in china. Sci Total Environ 802:149813
31. Benković-Lačić T, CulMak B, Benković R, Antunović S, Mirosavl Mević K (2022) Analysis of consumer opinions and habits related to apple consumption. In: Proceedings of the technique education agriculture management conference (10th International Scientific and Expert Conference TEAM2022), Slavonski Brod, Croatia, pp 455–459
32. Liang X, Zhang R, Gleason ML, Sun G (2022) Sustainable apple disease management in china: challenges and future directions for a trans-forming industry. Plant Dis 106(3):786–799
33. Rajan R, Ahmad MF, Singh J, Pandey K et al (2023) Organic apple production and prospects. In: Apples. CRC Press, pp 147–160
34. Jeyavishnu K, Thulasidharan D, Shereen MF, Arumugam A (2021) In-creased revenue with high value-added products from cashew apple (anacardium occidentale l.)—addressing global challenges. Food Bioprocess Technol 14:985–1012
35. Liu C-H, Chen T-L, Pai, Chiu C-H, Peng W-G, Weng C-C (2019) An intelligent robotic system for handling and laser marking fruits. In: Technologies and eco-innovation towards sustainability I: eco design of products and services, pp 75–88
36. Zhang Z, Zhang Z, Wang X, Liu H, Wang Y, Wang W (2019) Models for economic evaluation of multi-purpose apple harvest platform and soft-ware development. Int J Agric Biol Eng 12(1):74–83
37. Boini A, Casadio N, Bresilla K, Perulli GD, Manfrini L, Grappadelli LC, Morandi B (2022) Early apple fruit development under photo-selective nets. Sci Hortic 292:110619
38. Hyson DA (2011) A comprehensive review of apples and apple components and their relationship to human health. Adv Nutr 2(5):408–420
39. Zhu Z, Jia Z, Peng L, Chen Q, He L, Jiang Y, Ge S (2018) Life cycle assessment of conventional and organic apple production systems in china. J Clean Prod 201:156–168
40. Sparrow R, Howard M (2021) Robots in agriculture: prospects, impacts, ethics, and policy. Precis Agric 22:818–833
41. Singh S, Jain P (2022) Applications of artificial intelligence for the development of sustainable agriculture. In: Agro-biodiversity and agri-ecosystem management. Springer, pp 303–322
42. Porter A, Alhamid J, Mo C, Miller J, Iannelli J, Honegger M, Lichtensteiger L (2020) Analysis and design of an auxiliary catching arm for an apple picking robot. In: ASME international mechanical engineering congress and exposition, vol 84546, American Society of Mechanical Engineers, p V07AT07A011
43. Kang H, Zhou H, Wang X, Chen C (2020) Real-time fruit recognition and grasping estimation for robotic apple harvesting. Sensors 20(19):5670
44. Wei J, Yi D, Bo X, Guangyu C, Dean Z (2020) Adaptive variable parameter impedance control for apple harvesting robot compliant picking. Complexity 2020:1–15
45. Zhang K, Lammers K, Chu P, Li Z, Lu R (2021) System design and control of an apple harvesting robot. Mechatronics 79:102644

46. Chen W, Zhang J, Guo B, Wei Q, Zhu Z (2021) An apple detection method based on des-yolo v4 algorithm for harvesting robots in complex environment. Math Probl Eng 2021:1–12
47. Fei Z, Vougioukas SG (2021) Co-robotic harvest-aid platforms: Real-time control of picker lift heights to maximize harvesting efficiency. Comput Electron Agric 180:105894
48. Yu X, Fan Z, Wang X, Wan H, Wang P, Zeng X, Jia F (2021) A lab-customized autonomous humanoid apple harvesting robot. Comput Electr Eng 96:107459
49. Wu C, Wang Y, Sun Q, Zhao Y, Zhang L (2021) Design and test of auxiliary harvesting device of apple. Recent Pat Eng 15(1):107–116
50. Hu G, Chen C, Chen J, Sun L, Sugirbay A, Chen Y, Jin H, Zhang S, Bu L (2022) Simplified 4-dof manipulator for rapid robotic apple harvesting. Comput Electron Agric 199:107177
51. Krakhmalev O, Gataullin S, Boltachev E, Korchagin S, Blagoveshchensky I, Liang K (2022) Robotic complex for harvesting apple crops. Robotics 11(4):77
52. Chen K, Li T, Yan T, Xie F, Feng Q, Zhu Q, Zhao C (2022) A soft gripper design for apple harvesting with force feedback and fruit slip detection. Agriculture 12(11):1802
53. Zhang K, Lammers K, Chu P, Dickinson N, Li Z, Lu R (2022) Algorithm design and integration for a robotic apple harvesting system. In: 2022 IEEE/RSJ International Conference on Intelligent Robots and Systems (IROS), IEEE, pp 9217–9224
54. Xiong Z, Feng Q, Li T, Xie F, Liu C, Liu L, Guo X, Zhao C (2022) Dual-manipulator optimal design for apple robotic harvesting. Agronomy 12(12):3128
55. Yoshida T, Onishi Y, Kawahara T, Fukao T (2022) Automated harvesting by a dual-arm fruit harvesting robot. ROBOMECH J 9(1):1–14
56. Au W, Chen C, Liu T, Kok E, Wang X, Zhou H, Wang MY (2022) The Monash apple retrieving system, Available at SSRN 4272682
57. Zhang Z, Zhou J, Yi B, Zhang B, Wang K (2023) A flexible swallowing gripper for harvesting apples and its grasping force sensing model. Comput Electron Agric 204:107489
58. Wang X, Kang H, Zhou H, Au W, Wang MY, Chen C (2023) Develop-ment and evaluation of a robust soft robotic gripper for apple harvesting. Comput Electron Agric 204:107552
59. Zhou H, Kang H, Wang X, Au W, Wang MY, Chen C (2023) Branch interference sensing and handling by tactile enabled robotic apple harvesting. Agronomy 13(2):503
60. Bac CW, Van Henten EJ, Hemming J, Edan Y (2014) Harvesting robots for high-value crops: state-of-the-art review and challenges ahead. J Field Robot 31(6):888–911
61. Shewfelt RL, Prussia SE (2022) Challenges in handling fresh fruits and vegetables. In: Postharvest handling, Elsevier, pp 167–186
62. Zhou J, He L, Karkee M, Zhang Q (2016) Analysis of shaking-induced cherry fruit motion and damage. Biosys Eng 144:105–114
63. Sola-Guirado RR, Castro-Garcia S, Blanco-Roldán GL, Gil-Ribes JA, González-Sánchez EJ (2020) Performance evaluation of lateral canopy shakers with catch frame for continuous harvesting of oranges for juice industry. Int J Agric Biol Eng 13(3):88–93
64. Wang W, Lu H, Zhang S, Yang Z (2019) Damage caused by multiple im-pacts of litchi fruits during vibration harvesting. Comput Electron Agric 162:732–738
65. Pu Y, Toudeshki A, Ehsani R, Yang F, Abdulridha J (2018) Selection and experimental evaluation of shaking rods of canopy shaker to reduce tree damage for citrus mechanical harvesting. Int J Agric Biol Eng 11(2):48–54
66. Sanders K (2005) Orange harvesting systems review. Biosys Eng 90(2):115–125
67. Bac CW, Hemming J, Van Henten EJ (2014) Stem localization of sweet-pepper plants using the support wire as a visual cue. Comput Electron Agric 105:111–120
68. Bac CW, Hemming J, Van Tuijl B, Barth R, Wais E, van Henten EJ (2017) Performance evaluation of a harvesting robot for sweet pepper. J Field Robot 34(6):1123–1139
69. Lehnert C, Sa I, McCool C, Upcroft B, Perez T (2016) Sweet pepper pose detection and grasping for automated crop harvesting. In 2016 IEEE international conference on robotics and automation (ICRA), IEEE, pp 2428–2434
70. Lehnert C, English A, McCool C, Tow AW, Perez T (2017) Autonomous sweet pepper harvesting for protected cropping systems. IEEE Robot Autom Lett 2(2):872–879

71. Lehnert C, McCool C, Sa I, Perez T (2020) Performance improvements of a sweet pepper harvesting robot in protected cropping environments. J Field Robot 37(7):1197–1223
72. Arad B, Balendonck J, Barth R, Ben-Shahar O, Edan Y, Hell-stro¨m T, Hemming J, Kurtser P, Ringdahl O, Tielen T et al (2020) Development of a sweet pepper harvesting robot. J Field Robot 37(6):1027–1039
73. Fan P, Yan B, Wang M, Lei X, Liu Z, Yang F (2021) Three-finger grasp planning and experimental analysis of picking patterns for robotic apple harvesting. Comput Electron Agric 188:106353
74. Lu Y, Lu R, Zhang Z (2022) Development and preliminary evaluation of a new apple harvest assist and in-field sorting machine. Appl Eng Agric 38(1):23–35
75. Wu L, Ma J, Zhao Y, Liu H (2021) Apple detection in complex scene using the improved yolov4 model. Agronomy 11(3):476
76. Liu DW (2022) Hierarchical optimal path planning (hopp) for robotic apple harvesting
77. Gao R, Zhou Q, Cao S, Jiang Q (2022) An algorithm for calculating apple picking direction based on 3D vision. Agriculture 12(8):1170
78. Lv J, Wang Y, Xu L, Gu Y, Zou L, Yang B, Ma Z (2019) A method to obtain the near-large fruit from apple image in orchard for single-arm apple harvesting robot. Sci Hortic 257:108758
79. Forshey C, Elfving DC, Stebbins RL et al (1992) Training and pruning apple and pear trees. American Society for Horticultural Science
80. Jia W, Wang Z, Zhang Z, Yang X, Hou S, Zheng Y (2022) A fast and efficient green apple object detection model based on FoveaBox. J King Saud Univ-Comput Inf Sci 34(8):5156–5169
81. Gao F, Fu L, Zhang X, Majeed Y, Li R, Karkee M, Zhang Q (2020) Multi-class fruit-on-plant detection for apple in snap system using faster R-CNN. Comput Electron Agric 176:105634
82. Chu P, Li Z, Lammers K, Lu R, Liu X (2020) Deepapple: Deep learning-based apple detection using a suppression mask R-CNN. arXiv preprint arXiv:2010.09870
83. Mazzia V, Khaliq A, Salvetti F, Chiaberge M (2020) Real-time apple detection system using embedded systems with hardware accelerators: An edge AI application. IEEE Access 8:9102–9114
84. Li Q, Jia W, Sun M, Hou S, Zheng Y (2021) A novel green apple segmentation algorithm based on ensemble u-net under complex orchard environment. Comput Electron Agric 180:105900
85. Kang H, Chen C (2020) Fruit detection, segmentation and 3d visualisation of environments in apple orchards. Comput Electron Agric 171:105302
86. Kuznetsova A, Maleva T, Soloviev V (2020) Using yolov3 algorithm with pre-and post-processing for apple detection in fruit-harvesting robot. Agronomy 10(7):1016
87. Fu L, Majeed Y, Zhang X, Karkee M, Zhang Q (2020) Faster R–CNN–based apple detection in dense-foliage fruiting-wall trees using RGB and depth features for robotic harvesting. Biosys Eng 197:245–256
88. Jia W, Mou S, Wang J, Liu X, Zheng Y, Lian J, Zhao D (2020) Fruit recognition based on pulse coupled neural network and genetic Elman algorithm application in apple harvesting robot. Int J Adv Rob Syst 17(1):1729881419897473
89. Jiao Y, Luo R, Li Q, Deng X, Yin X, Ruan C, Jia W (2020) Detection and localization of overlapped fruits application in an apple harvesting robot. Electronics 9(6):1023
90. Jia W, Tian Y, Luo R, Zhang Z, Lian J, Zheng Y (2020) Detection and segmentation of overlapped fruits based on optimized mask R-CNN application in apple harvesting robot. Comput Electron Agric 172:105380
91. Kuznetsova A, Maleva T, Soloviev V (2021) Yolov5 versus yolov3 for apple detection. In: Cyber-physical systems: modelling and intelligent control. Springer, pp 349–358
92. Ji W, Pan Y, Xu B, Wang J (2022) A real-time apple targets detection method for picking robot based on shuflenetv2-YOLOX. Agriculture 12(6):856
93. Pugh A (2013) Robot vision. Springer Science & Business Media
94. Nasrabadi M, Kordrostami M, Ghasemi-Soloklui AA, Gararazhian M, Gharaghani A (2022) Regulations of form-training and pruning. In: Apples. CRC Press, pp 85–104
95. van Marrewijk BM, Vroegindeweij BA, Gené-Mola J, Mencarelli A, Hemming J, Mayer N, Wenger M, Kootstra G (2022) Evaluation of a boxwood topiary trimming robot. Biosyst Eng 214:11–27

96. Molina J, Hirai S (2017) Aerial pruning mechanism, initial real environment test. Robot Biomim 4(1):1–11
97. Zahid A, Mahmud MS, He L, Heinemann P, Choi D, Schupp J (2021) Technological advancements towards developing a robotic pruner for apple trees: a review. Comput Electron Agric 189:106383
98. He L, Schupp J (2018) Sensing and automation in pruning of apple trees: a review. Agronomy 8(10):211
99. Zeng H, Yang J, Yang N, Huang J, Long H, Chen Y (2022) A review of the research progress of pruning robots. In: 2022 IEEE 2nd International Conference on Data Science and Computer Application (ICDSCA), IEEE, pp 1069–1073
100. Saure MC (1987) Summer pruning effects in apple—a review. Sci Hortic 30(4):253–282
101. Wang Y, Zhang X, Xie L, Zhou J, Su H, Zhang B, Hu X (2020) Pruning from scratch. In: Proceedings of the AAAI conference on artificial intelligence, vol. 34, pp 12273–12280
102. Lehnert R (2012) Robotic pruning. Good Fruit Grower Nov. 1, 2012. https://www.goodfruit.com/robotic-pruning/
103. Li D, Wang P, Du L (2018) Path planning technologies for autonomous underwater vehicles-a review. IEEE Access 7:9745–9768
104. Karkee M, Adhikari B, Amatya S, Zhang Q (2014) Identification of pruning branches in tall spindle apple trees for automated pruning. Comput Electron Agric 103:127–135. https://doi.org/10.1016/j.compag.2014.02.013
105. Ji W, Qian Z, Xu B, Tao Y, Zhao D, Ding S (2016) Apple tree branch segmentation from images with small gray-level difference for agricultural harvesting robot. Optik 127(23):11173–11182. https://doi.org/10.1016/j.ijleo.2016.09.044
106. Zhang J, He L, Karkee M, Zhang Q, Zhang X, Gao Z (2018) Branch detection for apple trees trained in fruiting wall architecture using depth features and Regions-Convolutional Neural Network (R-CNN). Comput Electron Agric 155:386–393. https://doi.org/10.1016/j.compag.2018.10.029
107. Yang CH, Xiong LY, Wang Z, Wang Y, Shi G, Kuremot T, Zhao WH, Yang Y (2020) Integrated detection of citrus fruits and branches using a convolutional neural network. Comput Electron Agric 174. https://doi.org/10.1016/j.compag.2020.105469
108. Hashimoto K (2003) A review on vision-based control of robot manipulators. Adv Rob 17(10):969–991. https://doi.org/10.1163/156855303322554382
109. Gongal A, Silwal A, Amatya S, Karkee M, Zhang Q, Lewis K (2016) Apple crop-load estimation with over-the-row machine vision system. Comput Electron Agric 120:26–35. https://doi.org/10.1016/j.compag.2015.10.022
110. Majeed Y, Zhang J, Zhang X, Fu L, Karkee M, Zhang Q, Whiting MD (2020) Deep learning based segmentation for automated training of apple trees on trellis wires. Comput Electron Agric 170. https://doi.org/10.1016/j.compag.2020.105277
111. Zhang Y, He S, Wa S, Zong Z, Liu Y (2021) Using generative module and pruning inference for the fast and accurate detection of apple flower in natural environments. Information 12(12):495
112. Wang Q, Zhang Q (2013) Three-dimensional reconstruction of a dormant tree using RGB-D cameras. In: American society of agricultural and biological engineers, St. Joseph Paper number 131593521, pp 1. ASABE. https://doi.org/10.13031/aim.20131593521
113. Adhikari B (2012) Identification of pruning branches in tall spindle apple trees for automated pruning. Ph.D. thesis, Washington State University
114. You A, Sukkar F, Fitch R, Karkee M, Davidson JR (2020) An efficient planning and control framework for pruning fruit trees. In: 2020 IEEE International Conference on Robotics and Automation (ICRA), IEEE, pp 3930–3936
115. Korayem M, Shafei A, Seidi E (2014) Symbolic derivation of governing equations for dual-arm mobile manipulators used in fruit-picking and the pruning of tall trees. Comput Electron Agric 105:95–102
116. Zahid A, Mahmud MS, He L, Choi D, Heinemann P, Schupp J (2020) Development of an integrated 3R end-effector with a Cartesian manipulator for pruning apple trees. Comput Electron Agric 179. https://doi.org/10.1016/j.compag.2020.105837

117. Zahid A, He L, Zeng L, Choi D, Schupp J, Heinemann P (2020) Development of a robotic end-effector for apple tree pruning. Trans ASABE 63(4):847–856
118. You A, Kolano H, Parayil N, Grimm C, Davidson JR (2022) Precision fruit tree pruning using a learned hybrid vision/interaction controller. In: 2022 International Conference on Robotics and Automation (ICRA), IEEE, pp 2280–2286
119. Zahid A, Mahmud MS, He L, Schupp J, Choi D, Heinemann P (2022) An apple tree branch pruning analysis. HortTechnology 32(2):90–98
120. Karkee M, Adhikari B (2015) A method for three-dimensional reconstruction of apple trees for automated pruning. Trans ASABE 58(3):565–574
121. Akbar SA, Chattopadhyay S, Elfiky NM, Kak A (2016) A novel bench-mark RGBD dataset for dormant apple trees and its application to automatic pruning. In: Proceedings of the IEEE conference on computer vision and pattern recognition workshops, pp 81–88
122. Elfiky N (2022) Application of artificial intelligence in the food industry: AI-based automatic pruning of dormant apple trees. In: Artificial intelligence: a real opportunity in the food industry. Springer, pp 1–15
123. Karkee M, Adhikari B (2015) A method for three-dimensional reconstruction of apple trees for automated pruning. Trans ASABE 58(3):565–574. https://doi.org/10.13031/trans.58.10799
124. Díaz CA, Pérez DS, Miatello H, Bromberg F (2018) Grapevine buds detection and localization in 3D space based on structure from motion and 2D image classification. Comput Ind 99:303–312. https://doi.org/10.1016/j.compind.2018.03.033
125. Katyara S, Ficuciello F, Caldwell DG, Chen F, Siciliano B (2020) Reproducible pruning system on dynamic natural plants for field agricultural robots, pp 1–15. http://arxiv.org/abs/2008.11613
126. Baugher T, Jarvinen T, Dugan E, Schupp J (2016) Can a rules-based apple pruning system improve labor efficiency without affecting orchard productivity? PA Fruit News 96(2):16–17
127. Zahid A, He L, Zeng L, Choi D, Schupp J, Heinemann P (2020a) Development of a robotic end-effector for apple tree pruning. Trans ASABE 63(4):847–856. https://doi.org/10.13031/trans.13729
128. Kondo N, Ting KC (1998) Robotics for plant production. Artif Intell Rev 12(1–3):227–243. https://doi.org/10.1007/978-94-011-5048-4_12
129. Botterill T, Paulin S, Green R, Williams S, Lin J, Saxton V, Mills S, Chen XQ, Corbett-Davies S (2017) A robot system for pruning grape vines. J Field Rob 34(6):1100–1122. https://doi.org/10.1002/rob.21680
130. Huang B, Shao M, Chen W (2016) Design and research on end effector of a pruning robot. Int J Simulat—Syst Sci Technol 17(36):1–5. https://doi.org/10.5013/IJSSST.a.17.36.19
131. Dallabetta N, Forno F, Mattedi L, Giordan M, Wehrens H et al (2014) The implication of different pruining methods on apple training systems. POLJOPRIVREDA I SUMARSTVO 60(4):173–179
132. Jackson D, Looney N, Palmer J (2010) Pruning and training of deciduous fruit trees. In: Temperate and subtropical fruit production, CABI Wallingford UK, pp 44–61
133. Zahid A, He L, Choi D, Schupp J, Heinemann P (2021) Investigation of branch accessibility with a robotic pruner for pruning apple trees. Trans ASABE 64(5):1459–1474
134. Schupp JR, Winzeler HE, Kon TM, Marini RP, Baugher TA, Kime LF, Schupp MA (2017) A method for quantifying whole-tree pruning severity in mature tall spindle apple plantings. HortScience 52(9):1233–1240
135. Zhang X, He L, Majeed Y, Whiting MD, Karkee M, Zhang Q (2018) A precision pruning strategy for improving efficiency of vibratory mechanical harvesting of apples. Trans ASABE 61(5):1565–1576
136. Franzen J, Hirst P (2014) Optimal pruning of apple and effects on tree architecture, productivity, and fruit quality. In: XXIX International Horticultural Congress on horticulture: sustaining lives, livelihoods and landscapes (IHC2014), vol 1130, pp 307–310
137. Bai J, Xing H, Ma S, Wang M (2019) Studies on parameter extraction and pruning of tall-spindle apple trees based on 2D laser scanner. IFAC-Pap Online 52(30):349–354

138. Liu S, Yao J, Li H, Qiu C, Liu R (2019) Research on a method of fruit tree pruning based on BP neural network. J Phys Conf Ser, vol 1237. IOP Publishing, p 042047
139. Zahid A, He L, Choi DD, Schupp J, Heinemann P (2020) Collision free path planning of a robotic manipulator for pruning apple trees. In: 2020 ASABE annual international virtual meeting. American Society of Agricultural and Biological Engineers, p 1
140. Robinson T, Dominguez L, Acosta F (2014) Pruning strategy affects fruit size, yield and biennial bearing of 'gala' and 'honey crisp' apples. In: XXIX International Horticultural Congress on horticulture: sustaining lives, livelihoods and landscapes (IHC2014), vol 1130, pp 257–264
141. Zahid A, Mahmud MS, He L (2021) Evaluation of branch cutting torque requirements intended for robotic apple tree pruning. In: 2021 ASABE annual international virtual meeting. American Society of Agricultural and Biological Engineers, p 1
142. Li Y, Ma S, Ding Z, Li L, Xin Y, Su C (2022) Path planning of a robotic manipulator for pruning apple trees based on RRT-connect algorithm. In: 2022 ASABE annual international meeting. American Society of Agricultural and Biological Engineers, p 1
143. Greene D, Costa G (2012) Fruit thinning in pome-and stone-fruit: state of the art. In: EUFRIN thinning working group symposia, vol 998, pp 93–102
144. Yoder K, Peck G, Combs L, Byers R (2012) Using a pollen tube growth model to improve apple bloom thinning for organic production. In: II international organic fruit symposium, vol 1001, pp 207–214
145. Dennis FJ (2000) The history of fruit thinning. Plant Growth Regul 31:1–16
146. Ilie A, Hoza D, Oltenacu V et al (2016) A brief overview of hand and chemical thinning of apple fruit. Sci Pap Ser B Hortic 60:59–64
147. Batjer LP, Billingsley HD et al (1964) Apple thinning with chemical sprays
148. Childers NF (1959) Chemical fruit thinning of peach and apple. Rutgers University
149. Edgerton L (1972) Control of abscission of apples with emphasis on thinning and pre-harvest drop. In: Symposium on growth regulators in fruit production, vol 34, pp 333–344
150. Webster A (1992) Tree growth control and fruit thinning; possible alternatives to the use of plant growth regulators. In: II international symposium on integrated fruit production, vol 347, pp 149–162
151. Wertheim S (1997) Chemical thinning of deciduous fruit trees. In: VIII international symposium on plant bioregulation in fruit production, vol 463, pp 445–462
152. Bangerth F (2000) Abscission and thinning of young fruit and thier regulation by plant hormones and bioregulators. Plant Growth Regul 31:43–59
153. Dorigoni A, Lezzer P (2007) Chemical thinning of apple with new com-pounds. Erwerbs-Obstbau 49(3):93–96
154. Verma P, Sharma S, Sharma N, Chauhan N (2022) Review on crop load management in apple (malus x domestica borkh.). J Hortic Sci Biotechnol:1–23
155. Schupp J, Baugher TA, Miller S, Harsh R, Lesser K (2008) Mechanical thinning of peach and apple trees reduces labor input and increases fruit size. HortTechnology 18(4):660–670
156. Blanke M, Damerow L (2008) A novel device for precise and selective thinning in fruit crops to improve fruit quality. In: International symposium on application of precision agriculture for fruits and vegetables, vol 824, pp 275–280
157. Solomakhin AA, Blanke MM (2010) Mechanical flower thinning improves the fruit quality of apples. J Sci Food Agric 90(5):735–741
158. Hehnen D, Hanrahan I, Lewis K, McFerson J, Blanke M (2012) Mechanical flower thinning improves fruit quality of apples and promotes consistent bearing. Sci Hortic 134:241–244
159. McClure KA, Cline JA (2015) Mechanical blossom thinning of apples and influence on yield, fruit quality and spur leaf area. Can J Plant Sci 95(5):887–896
160. Hussain M, He L, Schupp J, Heinemann P (2022) Green fruit removal dynamics for development of robotic green fruit thinning end-effector. J ASABE 65(4):779–788
161. Qureshi A, Loh N, Kwon YM, Smith D, Gee T, Bachelor O, McCulloch J, Nejati M, Lim J, Green R et al (2023) Seeing the fruit for the leaves: towards automated apple fruitlet thinning, arXiv preprint arXiv:2302.09716

162. Basak A, Juraś I, Bialkowski P, Blanke M, Damerow L () Efficacy of mechanical thinning of apple in Poland. In: EUFRIN thinning working group symposia vol 1138, pp 75–82
163. Lordan J, Alins G, Avila G, Torres E, Carbó J, Bonany J, Alegre S (2018) Screening of eco-friendly thinning agents and adjusting mechanical thinning on 'gala', 'golden delicious' and 'fuji' apple trees, Sci Hortic 239:141–155
164. Bhattarai U, Bhusal S, Majeed Y, Karkee M (2020) Automatic blossom detection in apple trees using deep learning. IFAC-Pap Online 53(2):15810–15815
165. Mirbod O, Choi D, Heinemann PH, He L, Schupp JR (2021) In-field apple size and location tracking using machine vision to assist fruit thinning and harvest decision-making. In: 2021 ASABE annual international virtual meeting. American Society of Agricultural and Biological Engineers, p 1
166. Roquer-Beni L, Alins G, Arnan X, Boreux V, García D, Hambäck PA, Happe A-K, Klein A-M, Miñarro M, Mody K et al (2021) Management-dependent effects of pollinator functional diversity on apple pollination services: a response–effect trait approach. J Appl Ecol 58(12):2843–2853
167. Hussain M, He L, Schupp J, Lyons D, Heinemann P (2023) Green fruit segmentation and orientation estimation for robotic green fruit thinning of apples. Comput Electron Agric 207:107734
168. Ramírez F, Davenport TL (2013) Apple pollination: a review. Sci Hortic 162:188–203
169. Potts SG, Neumann P, Vaissière B, Vereecken NJ (2018) Robotic bees for crop pollination: why drones cannot replace biodiversity. Sci Total Environ 642:665–667
170. Eyles A, Close DC, Quarrell SR, Allen GR, Spurr CJ, Barry KM, Whiting MD, Gracie AJ (2022) Feasibility of mechanical pollination in tree fruit and nut crops: a review. Agronomy 12(5):1113
171. Wang T, Chen B, Zhang Z, Li H, Zhang M (2022) Applications of ma-chine vision in agricultural robot navigation: a review. Comput Electron Agric 198:107085
172. Mu X, He L (2022) An advanced Cartesian robotic system for precision apple crop load management. In: 2022 ASABE annual international meeting. American Society of Agricultural and Biological Engineers, p 1
173. Diaz Guzman S, Henspeter D, Taylor M, Duan S (2021) Drone pollination of flowering vegetation for agricultural applications. In: ASME international mechanical engineering congress and exposition, vol 85581. American Society of Mechanical Engineers, p V004T04A023
174. Mu X, He L, Heinemann P, Schupp J, Karkee M (2023) Mask R-CNN based apple flower detection and king flower identification for precision pollination. Smart Agric Technol 4:100151
175. Dias PA, Tabb A, Medeiros H (2018) Apple flower detection using deep convolutional networks. Comput Ind 99:17–28
176. Dias PA, Tabb A, Medeiros H (2018) Multispecies fruit flower detection using a refined semantic segmentation network. IEEE Robot Autom Lett 3(4):3003–3010
177. Yuan W, Choi D, Bolkas D, Heinemann PH, He L (2022) Sensitivity examination of YOLOV4 regarding test image distortion and training dataset attribute for apple flower bud classification. Int J Remote Sens 43(8):3106–3130
178. Sun K, Wang X, Liu S, Liu C (2021) Apple, peach, and pear flower detection using semantic segmentation network and shape constraint level set. Comput Electron Agric 185:106150
179. Tian Y, Yang G, Wang Z, Li E, Liang Z (2020) Instance segmentation of apple flowers using the improved mask R–CNN model. Biosys Eng 193:264–278
180. Sharma RR, Reddy S, Jhalegar M (2014) Pre-harvest fruit bagging: a useful approach for plant protection and improved post-harvest fruit quality—a review. J Hortic Sci Biotechnol 89(2):101–113
181. Yang H, Gu F, Wu F, Wang B, Shi L, Hu Z (2022) Production, use and recycling of fruit cultivating bags in china. Sustainability 14(21):14144
182. Ali M, Anwar R, Yousef A, Li B, Luvisi A, De Bellis L, Aprile A, Chen F (2021) Influence of bagging on the development and quality of fruits. Plants 10:358

183. Xu Y, Liu Y, Li W, Yang C, Lin Y, Wang Y, Chen C, Wan C, Chen J, Gan Z (2022) The effects of bagging on color change and chemical composition in 'jinyan' kiwifruit (*Actinidia chinensis*). Horticulturae 8(6):478
184. Ali MM, Anwar R, Yousef AF, Li B, Luvisi A, De Bellis L, Aprile A, Chen F (2021) Influence of bagging on the development and quality of fruits. Plants 10:358
185. Wang GP, Xue XM, Wang JZ (2021) Research progress and development trend of apple bagging technology in China. J Hebei Agric Sci 25:44–48
186. Kasso M, Bekele A (2018) Post-harvest loss and quality deterioration of horticultural crops in Dire Dawa Region, Ethiopia. J Saudi Soc Agric Sci 17(1):88–96
187. Gao H, Liu Y, Li D, Yu Y (2017) Vision localization algorithms for apple bagging robot. In: 2017 29th Chinese Control And Decision Conference (CCDC), IEEE, pp 135–140
188. Wang Y, Zhang Y, Pu Y, Zhang J, Wang F (2018) Design of a new fruit tree bagging machine. In: IOP conference series: materials science and engineering, vol 452. IOP Publishing, p 042099
189. Gou X, Zhang W, Zhang J, Zhang J, Zhang J (2019) Research on simulation and analysis of monitoring process of hail-proof apple bagging four-rotor aircraft. In: IOP conference series: materials science and engineering, vol 612. IOP Publishing, p 052030
190. Xia H, Zhen W, Chen D, Zeng W (2019) An ordinary multilayer fruit paper bag supplying device for fruit bagging. HortScience 54(9):1644–1649
191. Luo Z, Ma L, Zhou Z, Jia S, Fu Z (2019) Design and exploration of an apple sorting baler. In: IOP conference series: materials science and engineering, vol 612. IOP Publishing, p 032029
192. Xia H, Zhen W, Chen D, Zeng W (2020) Rigid-flexible coupling contact action simulation study of the open mechanism on the ordinary multi-layer fruit paper bag for fruit bagging. Comput Electron Agric 173:105414
193. Zhang W, Zhang F, Zhang J, Zhang J (2021) Kinematics analysis and trajectory planning computer simulation of smart apple bagging robot by hail suppression. J Phy Conf Ser 2033, IOP Publishing, p 012048
194. Liu X, Jia W, Ruan C, Zhao D, Gu Y, Chen W (2018) The recognition of apple fruits in plastic bags based on block classification. Precision Agric 19:735–749
195. Gou X, Zhang W, Zhang J, Zhang J, Zhang J (2019) Study on the structure design and feasibility analysis of apple inhaled box bags based on hailproof. In: IOP conference series: earth and environmental science, vol 252. IOP Publishing, p 052059
196. Zhang M, Zhang B, Li H, Shen M, Tian S, Zhang H, Ren X, Xing L, Zhao J (2020) Determination of bagged 'fuji' apple maturity by visible and near-infrared spectroscopy combined with a machine learning algorithm. Infrared Phys Technol 111:103529
197. Gou X, Zhang W, Zhang J, Zhang J, Zhang J (2020) Study on quality characteristics and feasibility analysis of hail-proof plastic bagging of 5000 mu in Gansu. In: IOP conference series: earth and environmental science, vol. 440. IOP Publishing, p 022048
198. Yuan L, Li Y, Cheng X, Ge S, Zhang Y (2022) Apple shockproof net cover automatic set machine design. In: Second international conference on Testing Technology and Automation Engineering (TTAE 2022), vol 12457, SPIE, pp 277–282
199. Pardo A, Borges PA (2020) Worldwide importance of insect pollination in apple orchards: a review. Agr Ecosyst Environ 293:106839
200. Garratt M, Breeze T, Boreux V, Fountain M, McKerchar M, Webber S, Coston D, Jenner N, Dean R, Westbury D et al (2016) Apple pollination: demand depends on variety and supply depends on pollinator identity. PLoS One 11(5):e0153889
201. Mu X, He L (2021) Mask R-CNN based king flowers identification for precise apple pollination. In: 2021 ASABE annual international virtual meeting. American Society of Agricultural and Biological Engineers, p 1

Chapter 2
Apple Bagging Technology Review and Design of a New End-Effector for Bagging Robot

Shahram Hamza Manzoor and Zhao Zhang

Abstract This book chapter gives an overview of current studies that have focused on the design and development of an apple bagging robotics system that incorporates detection and localization capabilities and a robust mechanical structure. This chapter presents key findings about sensor technologies like stereo vision and depth cameras and how they can be used to detect and locate apples in a complicated environment. Furthermore, it highlights current mechanical design and control developments, such as strategies for handling various apple shapes and sizes and optimizing overall system performance. This chapter also discusses future directions and challenges in this area, such as the need for more research on real-time sensor data processing, the development of more robust and reliable mechanical systems, and the integration of machine learning and other advanced algorithms to improve system performance.

Keywords Apple bagging · Stereo vision · Depth camera · Machine learning

S. H. Manzoor · Z. Zhang (✉)
Key Laboratory of Smart Agriculture System Integration, Ministry of Education, Beijing 100083, China
e-mail: zhaozhangcau@cau.edu.cn

Key Lab of Agricultural Information Acquisition Technology, Ministry of Agriculture and Rural Affairs, Beijing 100083, China

College of Information and Electrical Engineering, China Agricultural University, Beijing 100083, China

S. H. Manzoor
e-mail: shahramhamza786@uaar.edu.pk

Z. Zhang and X. Wang (eds.), *Towards Unmanned Apple Orchard Production Cycle*, Smart Agriculture 6, https://doi.org/10.1007/978-981-99-6124-5_2

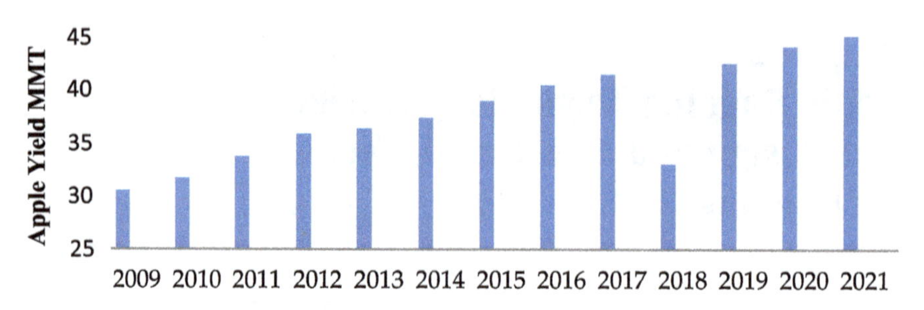

Fig. 2.1 China's apple production chart. *Source* National Bureau of Statistics (NBS), FAS Beijing

2.1 Introduction

2.1.1 The Apple Fruit Industry Importance in China

China is one of the world's leading apple fruit producers [1]. China's total apple fruit production is 20%, with a growing area of 30%, contributing significantly to the country's economy [2]. China accounts for more than 45% of global apple production. Figure 2.1 displays the annual apple fruit production from 2009 onward.

The following are the factors that contribute to the growing importance of China's apple exports:

- **Large Markets**: It is one of the largest producers and consumers of apples globally, with a significant proportion of its population relying on apples as a staple fruit. As an outcome, apples currently have a massive and expanding market locally and abroad.
- **Agricultural Importance**: Thousands of hectares of orchards are devoted to their cultivation. The apple industry is a major source of employment and income for many rural communities and a key contributor to the country's agricultural sector.
- **Food Security**: As a self-sufficient producer of apples, China can meet its demand for this fruit, which is considered an essential part of the national diet. That helps to ensure food security and stability for the country's population.
- **Cultural Significance**: Apples have a long history of usage and importance in China and are also regarded as a cultural emblem. They are often presented as presents on significant occasions and are frequently connected to health, wealth, and good fortune.

2.1.2 Major Regions Producing Apple Fruit in China

Based on Fig. 2.2, we observed that the brown area (Shandong and Shaanxi) displays apple production, accounting for 20% or more of apple products. In contrast, the proportion of the grey region in China's overall apple output is 5% or less.

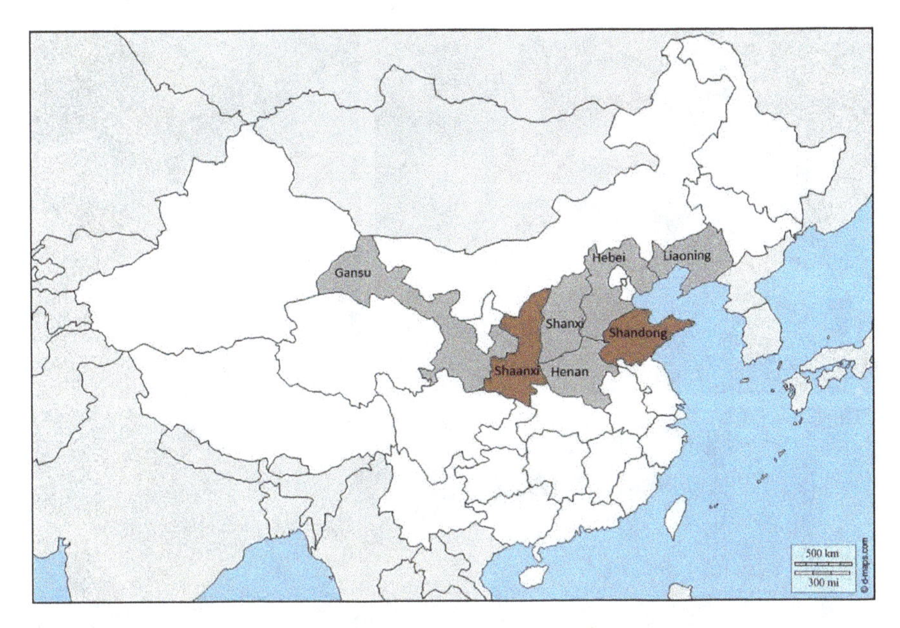

Fig. 2.2 Major provinces producing apple fruit

Table 2.1 presents key information on apple cultivation and production across various regions of China. Shaanxi emerges as a prominent player in this sector, boasting the largest area of planting (0.62 million hectares) and the highest apple production (0.66 million metric tons). Shandong, on the other hand, holds a significant position with its extensive planting area (0.25 million hectares) and substantial apple yield (0.55 million metric tons). These figures highlight the significant contributions of Shaanxi and Shandong to the overall apple production in the country.

Apple production in China is dominated by 'Fuji' varieties (Fig. 2.3). According to some sources, Fuji cultivars account for over 75% of China's entire crop. The majority of Fuji apple harvesting takes place in October. From July to September, the market is stocked with early-maturing cultivars like 'Gala' and 'Red Delicious'

Table 2.1 Major apple production with the cultivation area and their yield [3]

Regions	Area of planting (million hectares)	Apple production (million metric ton)
Gansu	0.24	0.20
Hebei	0.13	0.12
Henan	0.12	0.23
Liaoning	0.13	0.15
Shandong	0.25	0.55
Shanxi	0.15	0.23
Shaanxi	0.62	0.66

Fig. 2.3 Apples in Grocery Store (Left to Right Xinjiang, Aksu, Fuji, Guoguang) [4]

[4]. According to the reports, new cultivars like 'Venus and Gold' have been created by universities and private businesses for household growth. New varieties are rarely produced commercially because most farmers continue to choose Fuji cultivars due to their extended shelf life in cold storage.

According to the data presented in Fig. 2.4, China's apple imports were estimated to be 60,000 MT during the period of July 2021 to June 2022. The aforementioned statistics illustrate the noteworthy contribution made by China's apple fruit industry to the country's overall revenue. Between 1978 and 2022, there was a notable increase in China's apple cultivation area, overall production, per capita output, and yield. The production area has experienced a significant increase, expanding from 0.68 million hectares to 2.32 million hectares, with an average annual growth rate of 3.29%. The production of apples experienced a significant increase, rising from 2.28 metric tons to 60 metric tons. The comprehensive process of apple production is illustrated in Fig. 2.5. Figure 2.5(A) presents a timeline depicting the various orchard management practices that are typically carried out during a growing season. Figure 2.5(B) illustrates five specific practices, namely (1) the removal of fungal cankers from the trunk, (2) fruit thinning, (3) pesticide spraying in summer, (4) fruit debagging, and (5) fruit harvesting.

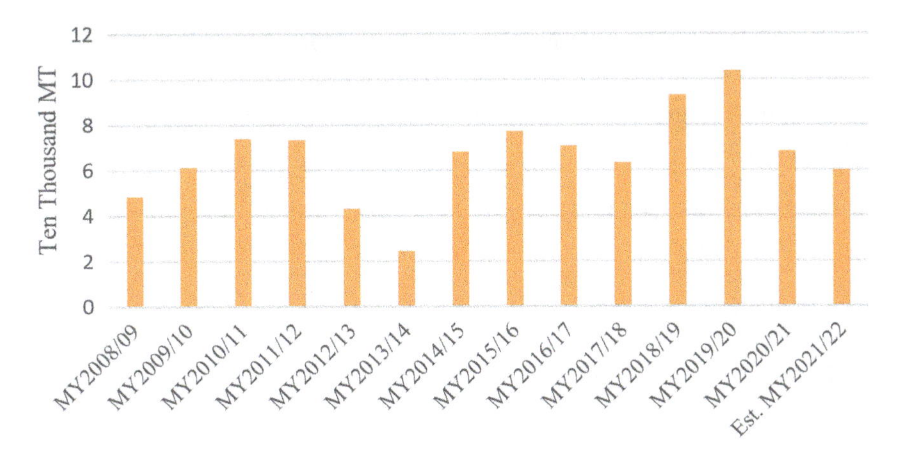

Fig. 2.4 China: Apple imports. *Source* China Customs

Fig. 2.5 Apple production process layout [3]

2.1.3 Apple Production Regulatory Authorities in China

In China, there are several government agencies and organizations that deal with apple production, including:

The Ministry of Agriculture and Rural Affairs: As far as promoting agricultural prosperity and rural progress goes, this organization reigns supreme amongst Chinese government bodies. One of their key responsibilities is developing policies and regulations that pertain to apple production. Apple cultivators also benefit significantly from the organization's aid.

The China National Apple Association: In China's apple industry, an esteemed non-profit organization serves as a key representative for producers seeking immense growth opportunities. Through its comprehensive service including but not limited to offering relevant market insights, technical expertise, and skills, and proactively engaging in policy advocacy efforts, this association ensures its members' overall satisfaction.

The China Fruit Marketing Association: A new addition to Chinas list of nonprofit organizations', focuses specifically on its thriving fruit industry. Its primary role involves gathering valuable market insights and disseminating them for the benefit of stakeholders at all levels. Beyond that, it supports industry growth initiatives while aiming for greater visibility and profitability for Chinese fruits in global markets.

The China Chamber of Commerce for Import and Export of Foodstuffs, Native Produce, and Animal By-Products: The aforementioned chamber of commerce operates under government affiliation and aids Chinese businesses that deal in the import and export of food and agricultural products, notably apples.

2.1.4 Major Apple Varieties in China

Step into a captivating adventure, exploring the world of apple variations in China. Every type of apple has a distinct tale, bridging us to diverse localities, customs, and flavors throughout China. By discovering their unique features, we develop a more profound acknowledgement for this humble and cherished fruit's incredible diversity.

The widespread appeal of apples stems from their beautiful hues, diverse flavors, and flexibility when it comes to using them in recipes—all factors that have earned them mass adoration worldwide. In this chapter, we introduce an assortment of various distinct types of apples, each with their regional popularity influenced by arresting traits found within, such as the Red Star apple (Fig. 2.6(a)), equipped with vivid red skin complementing juicy crisp flesh and growing extensively throughout China thanks to its sweet flavor, making it ideal not only as a fresh snack, but a cooking ingredient. Then, there's Golden Delicious (Fig. 2.6(b)), celebrated for its pulpy, golden-yellow, and enticingly moist flesh that allows its fruit to stand out among the rest. Not only is it sought after for snacking alone, but it is also a vital ingredient in making juices, cider, and similar apple-based culinary additives. Proceeding with our list, we come across the Fujian apple (Fig. 2.6(c)), which has a unique quality of being firm textured with juicy flesh whose sweeter tones better tenderize the palate when eaten for consumption or used recurrently as an essential ingredient for cooking. Last but foremost, is the Huaniu apple (Fig. 2.6(d)), traced back to Shaanxi province; known for exuding an intense sweetness balanced with sourness, boasting a sturdy consistency that makes it a perfect addition to dishes like baked goods. Amongst the world of apples lies one that reigns supreme with its irresistible taste—enter the brilliant Gala apple (Fig. 2.6(e)). This particular cultivar boasts sublime sweetness coupled with juicy pulp and an immensely satisfying crunchiness that wins fans all around the world. To round off this delightful journey exploring these divine fruits, we come upon the enchanting Snow apple (Fig. 2.6(f)). Its impossibly vibrant red skin holds within it flesh so perfectly balanced between sweet nectar-like juice and just enough acidity to keep things interesting.

These are just a few of the wide apple varieties grown in China, and the specific types developed can vary depending on the local climate and soil conditions. Despite the diversity of apple varieties, the country is still known for producing high-quality apples prized for their flavor, texture, and appearance.

Fig. 2.6 Apples of various types (a, b, c, d, e, and f)

2.1.5 Modernization in Apple Farming

In recent years, the state of apple production has been extensively studied by numerous researchers. For instance, Wang Hongye and Lu Xingkai selected 70 apple orchards that were representative of the Shaotong City region for their research study. The authors presented suggestions to improve the efficiency of apple cultivation in the frigid highlands of southwestern China, laying the foundation for industrial progress [2]. In their analysis of the current state of the Chinese apple industry, Song Zhe and Wang Hong identified the key challenges that the industry is facing and provided potential solutions for its future development [2]. Liu Xiaoguang, Zhou Ying, and their colleagues conducted a study comparing the international competitiveness of the apple industry in China and Japan. Based on their findings, they proposed a number of suggestions for improving competitive advantage. The details of their recommendations can be found in reference [3]. Zhai Heng and Shu Huairui have provided solutions to the challenges currently confronting the apple industry by analyzing the growth trajectory of the Chinese apple sector. Notwithstanding, fruit harvesting activities, which account for 50 to 70% of the aggregate labor hours, continue to rely on human effort. The automation of harvesting is expected in China due to the gradual decline of its farming workforce.

For instance, the utilization of a step ladder is imperative for bagging and harvesting operations, as the height of fruit trees poses a risk for manual harvesting, which is also less productive. Therefore, there is a strong incentive to automate and mechanize the process of harvesting. Mechanical harvesting experiments have been carried out in certain areas with the premise of single-pass harvesting, despite its limited usage. The harvesting technique that is frequently utilized necessitates the use of highly sophisticated robotic machinery [5]. In brief, it is imperative to develop an intelligent robot that possesses perceptual abilities akin to those of humans. The device is required to perform the tasks of fruit identification, location determination, and harvesting while ensuring the preservation of both the pericarp and the fruit tree.

2.1.6 Mechanization in Apple Farming: Improving Efficiency, Speed, and Accuracy

Apple production mechanization refers to using machines and automation technologies to assist in various aspects of apple production, from planting and harvesting to grading, packing, and storage. It can greatly increase efficiency, speed, and consistency in apple production while reducing the reliance on manual labor. Mechanization can be applied at many stages of the apple production process, including:

(a) **Planting**: Mechanized apple orchards often use planting machines to accurately plant trees in straight rows, reducing the time and labor required for manual planting, as shown in Fig. 2.7(a).

Fig. 2.7 Apple farming mechanization (a, b, c, d, e, and f)

(b) **Bagging**: Bagging apples as shown in Fig. 2.7(b) can help prevent damage from insect pests such as codling moths and diseases such as apple scabs.
(c) **Harvesting**: Apple harvesting may be challenging and labor-intensive, but robotics has significantly boosted its efficiency. When circumstances are favorable, automated apple harvesters may operate around the clock by picking the fruit with the help of sensors and mechanical arms, as shown in Fig. 2.7(c).
(d) **Sorting and Grading**: After harvesting, apples are sorted and graded according to size, color, and quality. Automated grading machines, as shown in Fig. 2.7(d): use cameras, sensors, and algorithms to accurately and quickly sort apples into different categories, reducing the need for manual inspection.
(e) **Packing and Bagging**: Automated apple packing and bagging systems, as shown in Fig. 2.7(e) can package apples into bags, boxes, or other containers much faster than manual labor. These systems often use sensors, conveyors, and other components to sort, bag, and label apples accurately and consistently.
(f) **Storage**: The improved performance of apple storage can be achieved through the utilization of automation technologies, as depicted in Fig. 2.7(f). Temperature and humidity monitoring systems are viable tools for monitoring the environmental conditions of apple storage facilities, thereby facilitating the preservation of the fruit's quality and freshness.

Mechanization in apple production has the potential to yield notable improvements in efficiency, productivity, and quality, while also resulting in decreased labor expenses. Nonetheless, the implementation of this approach necessitates the allocation of resources towards the acquisition of machinery and technology, and may not be viable for all cultivators or geographical areas.

2.1.7 *Apple Fruits Bagging Technologies and Their Importance of Apple Bagging*

The process of apple bagging is a crucial step in the production cycle of apples and can aid in multiple stages.

Protection: The process of bagging apples serves as a protective measure from different apple diseases and also prevents any potential damage, such as bruises or scratches, that may occur during the transportation and handling of the fruit.

Quality Control: Bagging apples is a practice that contributes to the preservation of the fruit's quality and freshness through quality control measures. Through meticulous regulation of the weight and dimensions of individual bags, cultivators can guarantee that the apples exhibit uniformity and superior caliber.

Branding and Marketing: The utilization of clear and visually appealing labeling on apple products has been found to contribute significantly to the promotion of the brand and the subsequent growth of sales. By utilizing labels, consumers can differentiate their apples from those of competitors by gaining additional knowledge about the variety of apple, the producer, and the weight of the fruit.

Transport and Storage: Bagged apples are a more convenient option for transportation and storage as compared to loose apples, owing to their reduced susceptibility to damage during handling. This practice aids in preserving the quality of the fruit, thereby mitigating the likelihood of wastage and spoilage.

Consumer Convenience: The aspect of consumer convenience is a significant factor as it pertains to the ease of portability, storage, and preparation of the product. The provision of products that are consistently of high quality can result in an increase in customer satisfaction for growers.

2.1.8 *Literature Review for Modernization of Apple Bagging*

2.1.8.1 Fruit Segmentation and Detection in Complex Orchard Environments Using Deep Learning, Color Index, and Image Processing Techniques

In [6], H. Gao, Y. Liu, D. Li, and Y. Yu et al. introduced a robot designed for apple bagging that utilizes vision localization techniques. The employed methodology integrates Otsu segmentation, enhanced connected component labeling, and a binocular stereo vision system to identify and locate young apples in the visible light spectrum. The circularity attribute is employed for the purpose of identifying and labeling apples. The findings demonstrate the efficacy and viability of the approach, as depicted in Fig. 2.8.

Researchers have made significant advances in fruit detection and segmentation in the pursuit of automating orchard harvesting systems and accurately assessing

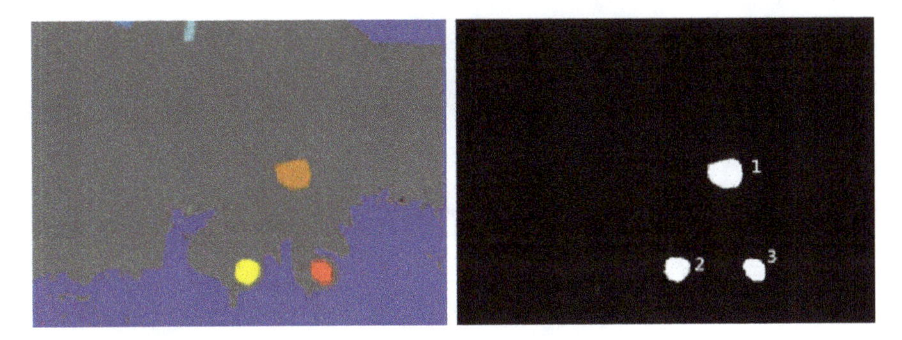

Fig. 2.8 The left image segmentation and recognition results [6]

yield. Reference [7] describes a ground-breaking study that introduces a segmentation model specifically designed for the difficult orchard environment. The proposed ensemble U-net architecture leverages edge structures and atrous convolutions to improve segmentation accuracy while preserving critical contextual information by leveraging the power of deep learning techniques. Furthermore, an atrous spatial pyramid pooling structure is used in the model to effectively combine features and improve generalization capabilities. The findings of this study show a significant improvement in fruit segmentation accuracy, paving the way for promising advances in evaluating orchard yield. Fig. 2.9 depicts the impact of the proposed methodology. Building on these achievements, J. Lv, F. Wang, and colleagues [8] developed a segmentation technique tailored specifically for images of bagged green apples, as shown in Fig. 2.10. This broadens the scope of fruit segmentation even further, catering to specific scenarios and providing valuable insights for orchard management and evaluation. The combination of these novel approaches has the potential to transform orchard practices and enable automated harvesting systems. The methodology entails the segregation of the fruit's conventional and illuminated areas. The Contrast Limited Adaptive Histogram Equalization (CLAHE) algorithm is employed for the purpose of enhancing edge definition. Additionally, the Red-Blue (R-B) color difference image is acquired to augment the color difference. The process of extracting the typical light region of the fruit involves the utilization of OTSU segmentation and denoising techniques. The region of interest is obtained through a process of subtracting the denoised reconstructed image from the initial image. The findings indicate that the method put forth is capable of proficiently segmenting images of green apples in bags, resulting in the acquisition of a comprehensive fruit region.

Zhang, Zou, and Pan (2019) developed an algorithm for the segmentation of apple fruits, as presented in reference [9], with the aim of achieving precision planting of apples. The study utilized three color and texture features that were extracted from the Grey-Level Co-occurrence Matrix (GLCM) to differentiate apple fruit pixels from other types of pixels. A Random Forest-based pixel classifier was devised, attaining a precision of 0.94. The algorithm underwent testing on a sample of 100 images depicting an apple orchard. The results indicated a segmentation error of 0.07, a false

Fig. 2.9 Segmentation effect of the method [7]

Fig. 2.10 The extraction of fruit area oriented images using normal light [8]: a, b, c, d, e, and f

positive rate of 0.13, and a false negative rate of 0.15. The algorithm proficiently partitions apple fruit in orchard images and can furnish a point of reference for accurate apple cultivation administration, as illustrated in Fig 2.13. [10] The present study introduces a novel approach for segmenting pixel patches in order to achieve precise identification of apples within natural orchard settings. This method is based on the grey-centre RGB color space. The approach utilizes chromatic characteristics and regional fluctuations within apple imagery to differentiate apple pixels from non-apple pixels, as illustrated in Fig 2.11. The method in question demonstrated a mean accuracy rate of 99.26% in comparison to alternative approaches.

K. Zou et al. [11] in (2011) developed a segmentation technique for apple images that relies on color indices. This method is intended for use by robots designed for harvesting and spraying tasks. The approach utilized in the study successfully partitioned apple images with high efficiency, yielding a mean pixel segmentation accuracy of 0.90 and a mean segmentation time of 20 ms. Jia et al. (2019) introduced a sturdy segmentation network framework in [12] for the purpose of detecting and segmenting fruits in natural orchards. The approach employs the Mask R-CNN model along with a Gaussian non-local attention mechanism to refine semantic features and accurately segment apples in intricate surroundings, as depicted in Fig 2.12. The findings indicate that the approach employed by the researchers surpasses other contemporary models, as evidenced by the AP box and AP mask metric values of 85.6 and 86.2%, respectively.

Fig. 2.11 Visualizing the outcomes of segmentation [10]: **a** light and thick shadows, **b** variable degrees of shadows and light

2.1.8.2 Deep Learning-Based Detection and Recognition of Fruit for Improved Yield Estimation in Orchards

Deep Learning and YOLO detection models have changed the way orchard yield prediction is done. In this significant study, researchers employed innovative techniques to identify and quantify mature Dezful native oranges in a vast orchard [13]. The top-performing YOLO-V4 model outperformed the competition in terms of precision (91.23%), recall (92.8%), F1-score (92%), and mAP (90.8%). The method's effectiveness is demonstrated in Fig. 2.14, which provides a practical solution for yield identification and measurement. While innovative imaging approaches addressed thin and dense canopies, spatial distribution mapping revealed a slight inaccuracy of +9.19% in fruit yield variations. These advancements empower orchard management by allowing for more informed decisions and increasing productivity.

In order to optimize vineyard harvesting processes, researchers created a computer vision algorithm, which is detailed in the paper [14]. The algorithm is divided into three steps. The proposed methodology involves a three-step process for grape cluster segmentation. Firstly, k-means clustering and color information are utilized to segment grape clusters. Secondly, edge images are extracted, and contour intersection points are determined. Finally, a geometric constraint method is employed to determine cutting points on the peduncle. The proposed methodology was subjected to experimentation using a dataset of 30 vineyard images. The results indicate an average recognition accuracy of 88.33%, with a success rate of 81.66% in visually detecting the cutting locations on the peduncles. The aforementioned discoveries underscore the capability of the approach to be employed by automated harvesting

Fig. 2.12 Different test images segmented using RS-Net, Retina Mask, YOLACT, and YOLACT were visualized [12]

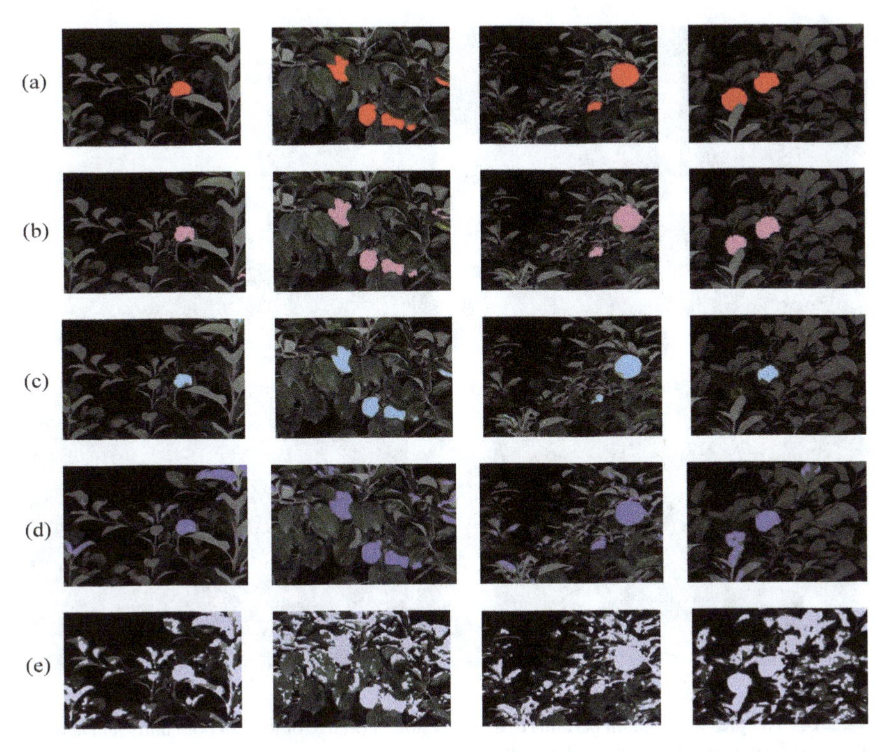

Fig. 2.13 The process of manually assigning labels to outcomes and the segmentation outcomes generated by an algorithm have been discussed in reference [9]. **a** The process of manually assigning labels to outcomes, **b** The segmentation approach that has been devised produces the following outcomes. **c** The Otsu method utilizing red-blue color space and boundary object removal, **d** The K-means clustering segmentation technique utilizing red-blue color space, **e** The adaptive threshold segmentation technique utilizing red-blue color space

machines, thereby furnishing a beneficial resolution in intricate vineyard settings. A thorough investigation was carried out in the paper denoted as [15] to devise a precise detection methodology for the purpose of tackling apple waste before thinning. The experts leveraged the YOLO V5's deep learning algorithm, which was further optimized through the utilization of channel pruning techniques. The model underwent pruning and fine-tuning, leading to the efficient and dependable detection of apple fruitlets as depicted in Fig. 2.15. The YOLO V5s model, which underwent channel pruning, exhibited commendable performance metrics. Specifically, it achieved a recall of 87.6%, precision of 95.8%, an F1 score of 91.5%, and a false detection rate of 4.2%. In addition, it is noteworthy that the model's size is 1.4 megabytes, and the mean duration for detecting each image was a mere 8 milliseconds. Researchers X. Liu, D. Zhao, W. Jia, W. Ji, and Y. Sun et al. introduced a machine vision approach to effectively recognize apples with partially red skins and varying tints, including green and pale yellow, in an important study referred to as [16]. The method under consideration employs Simple Linear Iterative Clustering (SLIC) to partition images

Fig. 2.14 The present research pertains to the detection of orange objects in both daily and at night images, utilizing the YOLO V3 and V4 models [13]

into super-pixel blocks, thereby enabling the extraction of color information and identification of potential regions. The utilization of a Histogram of Oriented Gradients (HOG) is a common practice in the field of computer vision to effectively capture and represent the shape of fruits, thereby enhancing the accuracy of their detection and localization. The method proposed has yielded noteworthy outcomes in terms of average recall, precision, and F1 scores, with values of 89.80%, 95.12%, and 92.38%, respectively. The results substantiate the proficiency of the methodology in precisely detecting and classifying partially red apples, thereby facilitating the advancement of fruit identification and quality evaluation in the realm of agriculture. P. Fan et al. presented a novel multi-feature patch-based apple image segmentation technique in their study [17]. The method enables fast and accurate detection of apple targets in apple-picking robots by utilizing the grey-centered RGB color space and a generalized K-means clustering algorithm. The method outperformed traditional segmentation methods and clustering algorithms on 240 apple photos, achieving impressive average accuracy (98.79%), a recall (99.91%), an F1 measure (99.35%), a false positive rate (0.04%), and false negative rate (1.18%). A system for detecting apple objects in orchard settings for picking robots was presented in a separate work in study [18].The method, which was based on Shufflenetv2-YOLOX with convolutional block attention module (CBAM) and adaptive spatial feature fusion (ASFF) modules, had a precision of 95.62%, a recall of 93.75%, and an F1 score of 0.95. The

Fig. 2.15 Pre-fruit thinning detection examples for apple fruitlets. **a** The outcome of a detection on an image taken in bright sunshine. Image recorded in direct sunlight and shot from behind **b** detection outcome. Image obtained in both cloudy and bright sunlight, with the resulting detection result shown in (**c**). Image acquired in overcast, backlit settings; the resulting detection result is shown in (**d**). Apple fruitlets were detected at a rate of 1, 3, and 5 under low, medium, and high light conditions, respectively. A single apple fruitlet has a detection result of 2, and misdetections equal 9. Occluded apple fruitlets (4), clustered apple fruitlets (7), apple fruitlets (8), and hazy apple fruitlets (9) are indicated by the detection findings 4, 6, 7, and 8 [15]

technique outperformed other lightweight networks in terms of detection accuracy and speed, operating at a detection rate of 65 frames per second. Furthermore, in [19], W. Ji, X. Gao, B. Xu, Y. Pan, and Z. Zhang presented an improved YOLOv4-based apple detection system. The method achieved an average precision of 93.42%, recall rate of 87.64%, and F1 value of 0.9035 by incorporating EfficientNet-B0 feature extraction and PANet networks. The improved system reduced storage memory by 87.8% while increasing recognition speed by 43% when compared to the original YOLOv4. Despite its smaller size and faster identification, Mobilenetv3-YOLOv4 accuracy remained comparable.

2.1.8.3 Apple Recognition and Tree Organ Classification Using Watershed Algorithm and Support Vector Machine

[20] This study proposed a method for 3D apple tree organ classification and yield estimation using a combination of color and shape features, which were fused and analyzed using a Support Vector Machine (SVM) algorithm with a linear kernel function. The method was tested on a dataset and outperformed other algorithms (KNN and Ensemble), achieving a recall of 93.75%, precision of 96.15%, and an F1 score

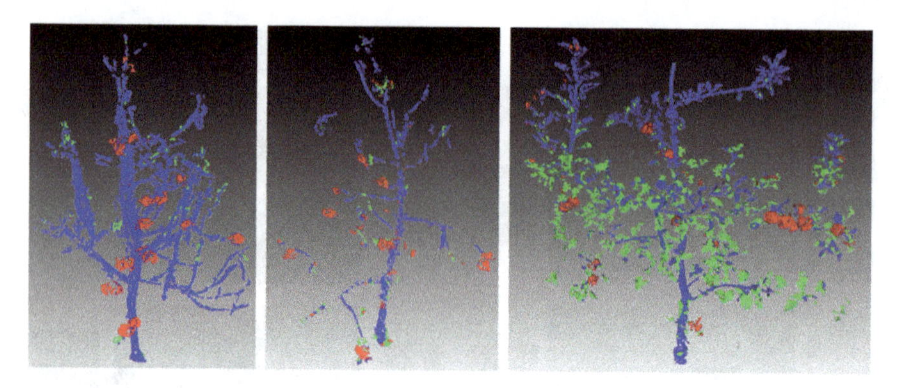

Fig. 2.16 Apple tree organs classification results before and after smoothing [20]

of 94.93%. The proposed method could be applied in real-world scenarios for automatic apple picking, pruning, and information management in orchards, as shown in Fig. 2.16. In [21], a block classification method based on the watershed algorithm and support vector machine for apple recognition in plastic bags was introduced. It segments the original images into irregular blocks using R-G grayscale edge detection. It classifies them into fruit and non-fruit using SVM based on color and texture features. The method effectively reduced the interference of light and improved recognition accuracy with a lower FNR and FPR compared to pixel classification.

2.1.8.4 Robotics for Apple Bagging

In [22], a new fruit tree bagging machine was designed (Fig. 2.17) to assist manual bagging and improve production efficiency. It considers the growth characteristics of different fruits and adjusts the spatial layout based on the steps and actions of manual bagging. W. J. Zhang et al. [23] designed apple an anti-hail bagging robot (Fig. 2.18) to reduce the labor intensity of fruit farmers, ensure a timely harvest, and improve the quality of fruits. It was tested in an orchard, demonstrating its capability to carry out crucial operations, including mobile navigation, picking, manipulation, and autonomous fruit packaging. The algorithms used by the robot were reliable and real-time enough to suit the needs for anti-hail apple bagging.

In [24], W. J. Zhang, F. Zhang, J. Zhang, and J. Zhang et al. concentrated on utilizing a supply device to open a multilayer fruit paper bag without damaging it (Fig. 2.19). The outcomes demonstrated that the slider stroke and speed of the open mechanism are crucial elements for the consequences of the bag opening. The researchers constructed a simulation model to analyze the effects. They found that a larger slider stroke can cause high rates of bag breakage, while a larger slider speed improves bag opening efficiency. The results of the simulation and experiments help improve the performance of the supplying device. H. Xia et al. [25] proposed a new supplying device for taking out and opening fruit paper bags. The device operates like

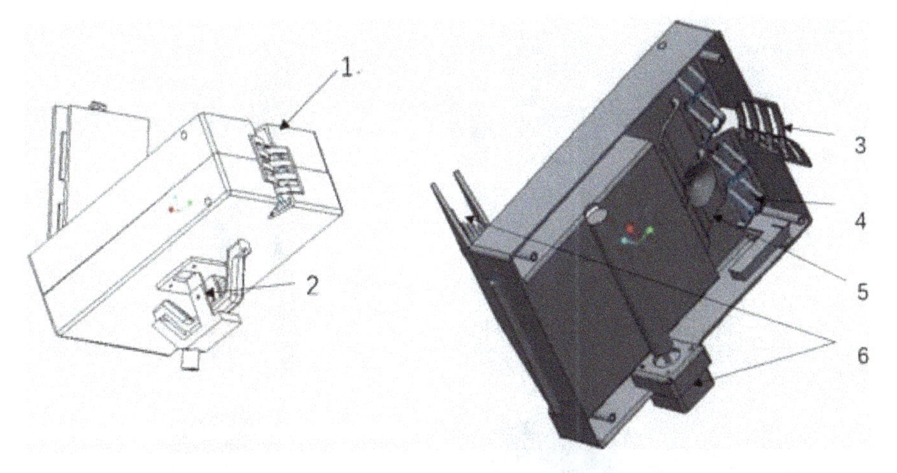

Fig. 2.17 Sketch of the fruit tree bagging device [22]: (1) The branches. (2) Handle and telescopic rod conversion device. (3) The leaf splitter. (4) Fruit bag folding and sealing device. (5) Fruit bag mouth device. (6) The fruit bag is pressed tightly out of the device

Fig. 2.18 Apple anti-hail apple bagging robot (a), and Apple Bagging Unmanned Quadrotor [23]

a farmer's hand, continuously taking and opening the bags from the inside. Laboratory experiments showed that the driving trajectory and speed significantly affected the operation's success rate and time. The device performed with an over 90% success rate and a 2-second opening time. In [26], a semi-automatic fruit bagging apparatus has been developed to address the challenges of labor-intensive and inefficient current methods (Fig. 2.20). The device is economical, easy to use, and effective since it delivers and releases paper bags one at a time using a self-locking mechanism.

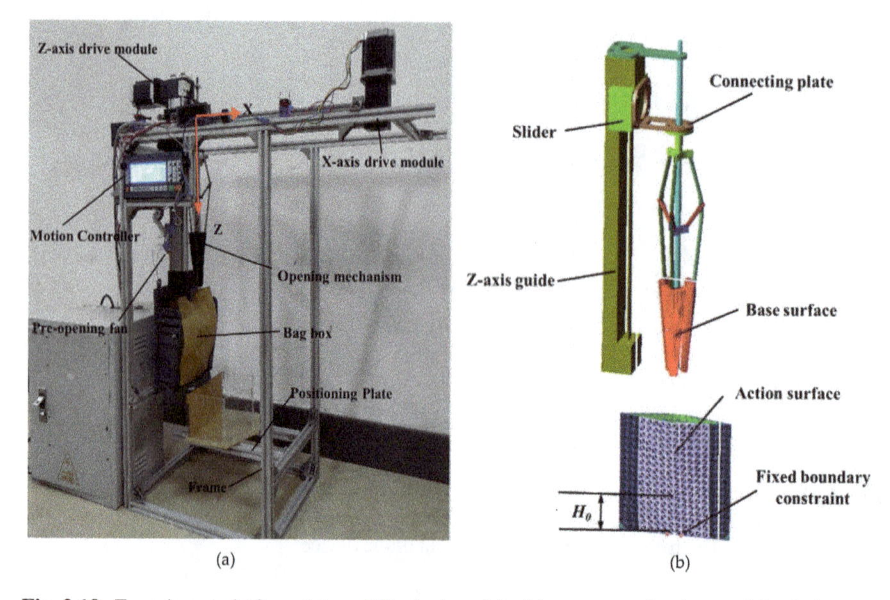

Fig. 2.19 Experiment platform (**a**), and Contact model of the open mechanism and the fruit paper bag (**b**) [24]

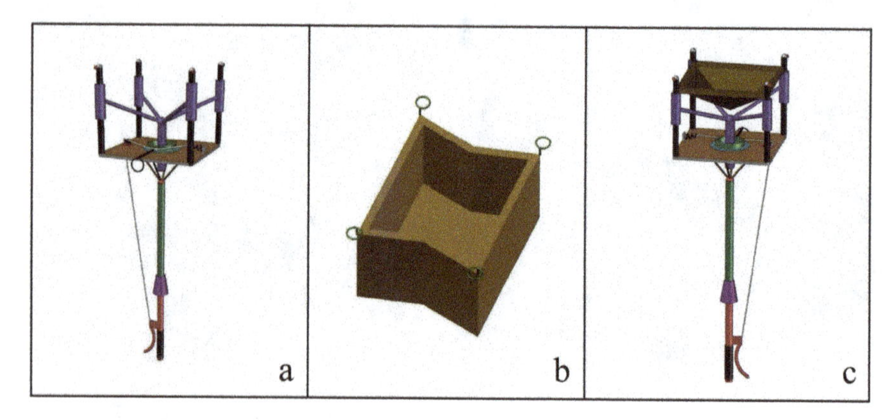

Fig. 2.20 The 3D design of the fruit bagging apparatus [26]: **a** fruit bagging apparatus; **b** paper bag; **c** fruit bagging apparatus with a paper bag

2.2 Design Requirements for Apple Bagging Robotics

In this section, there will be an in-depth discussion of what we have learned from the previous section of the literature review of related technologies. It will also include the process, hardware requirements, and the design process for apple bagging robotics. This section will also address the shortcomings of the previous studies. The design requirement for apple bagging is divided into modules or phases for better understanding.

The phase of the development:

2.2.1 (A) System Model for Apple Detection and Localization

- The computing machines: As the detection of apple fruit will form the basis of automated bagging, it will need a high-end GPU-equipped computing machine to process all the images in the repository.
- Data gathering: The process of fruit detection will be efficient and precise only if there is vast data to make a machine learn every pattern and detail in the form of images.
- To create an image database that is sufficiently vast to train the machine for accurate detection, these images of fruits should be taken in the morning, the evening, in the shade, in a minor shadow, and under low light.
- Image segmentation techniques are implemented in many studies in the literature review, thus making it an important process in the detection of apple fruit in Fig. 2.22. This is also a crucial knowledge area, and its in-depth practice is important (Fig. 2.21).
- The computer vision models for real-time detection: YOLO, YOLOv2, YOLOv3, YOLOv4, YOLOv5, YOLOv6, and now YOLOv7 and YOLOv8. Their knowledge and implementation are also important.
- Possible to customize new computer vision models for multiple tasks to improve apple fruit detection efficiency.
- Finding the discovered apple will be required after the detection procedure. Combining a detection model with a depth camera sensor is essential for achieving the objective.For the location and detection of apple fruit in space, this work will require [x-axis, y-axis, and z-axis] 3D information for a robotic arm.

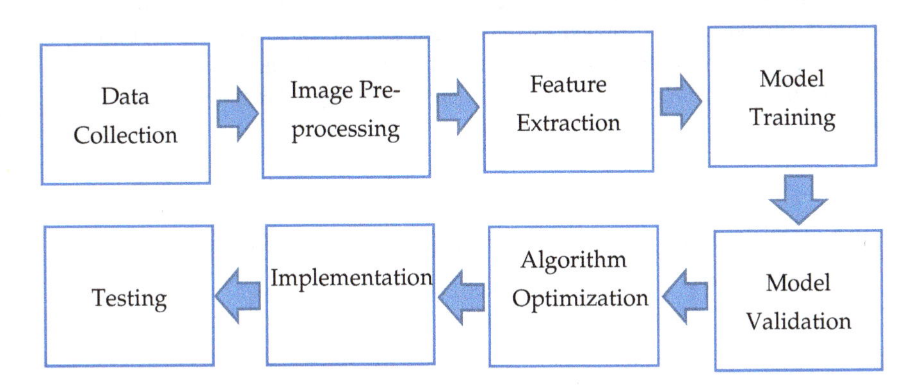

Fig. 2.21 System model for apple detection and localization

Fig. 2.22 Apple image segmentation. *Source* Quora

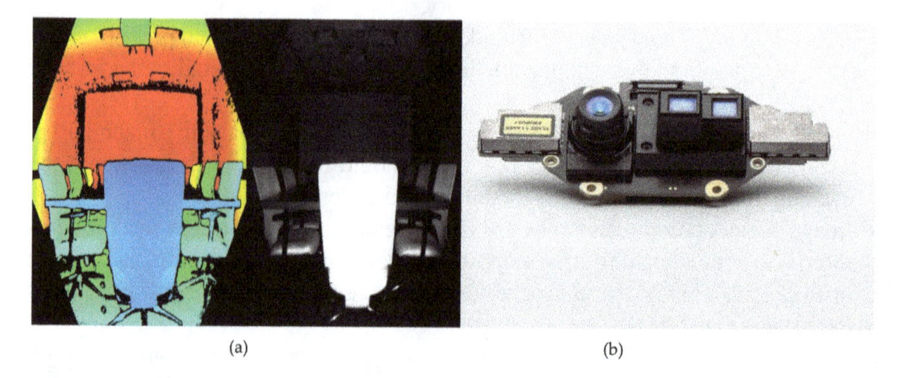

(a) (b)

Fig. 2.23 **a** Picture taken from the Azure Kinect depth camera **b** Azure depth camera. *Source* Microsoft

- To obtain 3D coordinates, a depth sensor-based camera such as the Azure Kinect depth camera isrequired (see (Figs. 2.23a, and b).

The host PC receives raw, modulated infrared (IR) images from the depth camera. The PC's GPU-accelerated depth engine software creates depth maps from the raw data. The depth camera supports multiple modes. Scenes with lower extents in the X and Y dimensions but bigger extents in the Z dimension are best suited for the narrow field of view (FoV) modes. The broad FoV settings work best when the scene contains huge X- and Y-extents but smaller Z-ranges, as shown in Fig. 2.23(a) (Fig. 2.24).

2.2.2 (B) System Model for Apple Bagging Mechanical Structure

- In most of the previous studies, the major focus was on the precise detection of apple fruit on a tree there was less work in building an electro-mechanical work integrated with these detection algorithms.

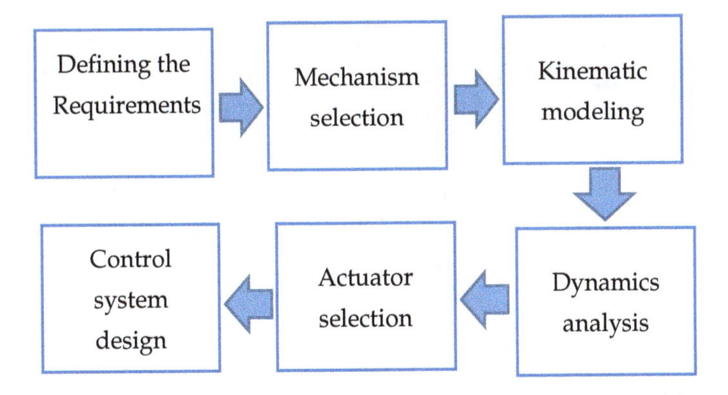

Fig. 2.24 System model for apple bagging mechanical structure

- The need to provide advanced mechanical design and determine the degree of freedom (DOF) for the robotic bagging platforms.
- 3D modeling expertise will be required for designing the frame of bagging robotics.
- A 3D printer will be needed for printing the frame parts and gears.
- Select the actuators, such as motors or pneumatic cylinders, based on the required torques and the operating conditions.
- Microcontroller intercommunication for parallel programming reduces these time lags and makes the system faster in execution.
- Flex sensor installation for effective and soft grip monitoring so that apple fruits are not damaged. Limiting switches could be installed as well for gripping.
- Every Electro-mechanical system requires a steady energy source/power supply. These modules will power up the system, including detection and localization.
- Design a user interface to control the arm and monitor its performance.
- Designing the advanced robotic arms (see Fig. 2.25).
- Initial laboratory testing of apple fruit bagging robotics and compilation of results.

2.2.3 *Evaluation and Optimization of Apple Fruit Bagging Robotics: Laboratory and Field Testing Report*

- This phase will analyze the initial laboratory tests and compile the report for improvements needed in the design.
- After making the improvements, this system will be tested again until there is enough satisfaction for further testing in the field.
- Initial field testing of apple fruit bagging robotics and compilation of results.

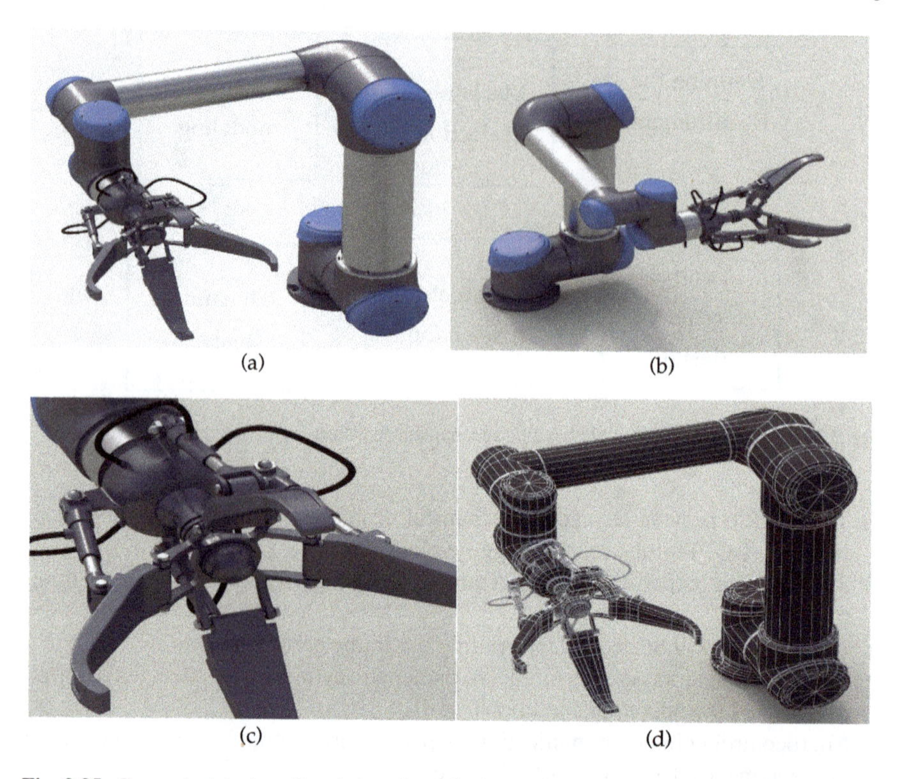

Fig. 2.25 Conceptual design of apple bagging robotics (a, b, c, and d). *Source* Free 3D modeling

2.2.4 *Enhancing the Design of Apple Fruit Bagging Robotics: Field Testing Analysis and Documentation*

- This phase will analyze the results of the initial field tests and compile a report on improvements needed in the design.
- After making improvements, this system will be tested again.
- The documentation process will be started based on compiled laboratory and field testing results.

2.2.5 *Product Delivery*

Documentation and product delivery needs to be improved.

2.2.6 Some Add-Ons for Agriculture Researchers

Some useful sensors could be integrated with this apple-bagging robotics and help the researcher with the data acquired in the field.

- GPS coordinates with the GPS module
- RFID sensor
- Apple starch content sensor
- Temperature sensor for temperature monitoring of apples during harvesting
- Weight sensor for yield estimation
- PH sensor

2.2.7 Benefits of Selecting Robotics Over Manual Bagging

Increased Efficiency: The utilization of Apple bagging robots results in improved productivity as they are capable of functioning incessantly, without any interruptions, and at a significantly accelerated rate in comparison to their human counterparts. This phenomenon can lead to a notable enhancement in productivity, enabling enterprises to manufacture a greater quantity of apple bags within a reduced time frame.

Improved Accuracy: The apple bagging robots are outfitted with advanced sensory and imaging technology that enables precise detection and measurement of the size, weight, and quality attributes of apples. As a consequence, the bagging process exhibits enhanced precision, whereby apples are systematically placed within bags with greater uniformity and at the appropriate mass.

Reduced Labor Costs: The automation of the bagging process can lead to a reduction in manual labor requirements, thereby resulting in substantial cost savings for companies. Furthermore, robots have the capability to operate in settings that could pose potential risks to human laborers.

Improved Safety: The Apple bagging robots have been engineered to ensure enhanced safety measures, incorporating sensors and safety mechanisms that serve to safeguard both workers and other personnel. On the contrary, the process of manually bagging items can be arduous and potentially hazardous to laborers.

Consistency: The utilization of apple bagging robots ensures a high level of consistency in the bagging process, as they are capable of maintaining a uniform speed and accuracy despite any potential discrepancies in apple size or quality. Achieving consistency and uniformity in the final product holds significant importance for customers.

2.2.8 *Disadvantages of Apple Bagging Robotics*

The adoption of robotics in apple fruit farming presents various benefits, including heightened efficiency, speed, and precision. However, it is essential to acknowledge several drawbacks associated with this approach.

High Cost: One of the primary drawbacks associated with the utilization of apple bagging robotics pertains to the elevated expenses associated with the technology. The considerable cost of acquiring robots, sensors, cameras, and other related components may pose a hindrance to certain apple bagging operations.

Maintenance and Repair: Maintenance and repair are necessary to ensure the proper functioning of the equipment. Regular attention is required to maintain optimal performance. The process can be both resource-intensive and financially demanding, necessitating expertise in specialized technical domains.

Limited Flexibility: The apple bagging robots exhibit restricted flexibility in comparison to human workers, despite their ability to be programmed for the handling of diverse apple varieties and sizes. In the event that an apple bagging enterprise alters the variety of apples that they are managing, it may be necessary to reprogram the robots, a process that can consume a significant amount of time.

Job Losses: The implementation of apple bagging robots may lead to a reduction in employment opportunities for human workers. As the bagging process becomes fully automated, certain employees may undergo retraining to operate alongside the robots, while others may face unemployment.

Dependence on Technology: The reliance on technology is evident in the case of Apple's automated bagging system, which is dependent on technological mechanisms for its proper functioning. The malfunctioning of robots can result in both time and financial inefficiencies.

2.2.9 *Challenges and Our Future Perspectives*

Complexities of Apple Detection and Localization: Apples deteaction and the research for reliable and efficient algorithms for detecting and localizing apples presents a number of issues that must be carefully considered. One critical factor is data availability, as machine learning and deep learning algorithms rely substantially on a large and impartial dataset to learn and detect apples with precision. Inadequacies or biases in the training data, on the other hand, can have a direct impact on the efficacy of these algorithms. The intricate backdrops in which fruits frequently find themselves contribute to the complexity of apple detection. Foliage, branches, and other nearby items might make it difficult to differentiate the fruit from its surroundings. Furthermore, changes in lighting circumstances complicate matters further, as apples may appear differently under different lighting configurations, making it difficult to precisely identify and detect them. Occlusion is another barrier to recognizing and localizing apples. Fruits hidden from view can defy detection by algorithms, whether partially or completely occluded by other objects, necessitating novel ways to address this difficulty. Furthermore, the diversity in apple shape, size, and color needs the development of comprehensive algorithms capable of handling their complexity. The detecting issue gets complicated further by scale changes, with apples appearing at various sizes and distances. Small or distant fruits are readily overlooked, and algorithms must be devised to account for these changes in order to assure complete detection. Furthermore, the position or stance of apples can fluctuate, making it challenging for algorithms to precisely recognize them. The ability to adapt to these position fluctuations is critical for successful detection and localization. The computational complexity of machine learning systems, particularly deep learning, adds another stumbling block to real-time apple identification. To ensure rapid and effective apple detection in real-world circumstances, extensive computational resources and optimization approaches are required. In apple identification, generalization, a major feature of machine learning, is critical. The wide range of physical features found in apples makes it difficult for algorithms to generalize their detection abilities to new, previously unknown data. To recognize apples properly in real-world contexts, algorithms must navigate environmental changes such as lighting differences, occlusions, and posture variations.

Mechanial Limitations: Choosing the mechanism that best fits the requirements, such as a serial, parallel, or hybrid. In most of the previous studies, serial programming for instruction execution was implemented. For example, if there are 5 motors embedded, and a command is generated to execute the instruction of motion for the 5 motors and it will be from motor 1 to motor 5 in a series flow. This takes a lot of time because motors are attracted to gears, and their frictional forces also add up to lag time. For the time required for motor 5 to do a certain movement, then, it will be [$\mathbf{T_5 = T_{GF5} + T_4 + T_{GF4} + T_3 + T_{GF3} + T_2 + T_{GF2} + T_1 + T_{GF1}}$]. This time calculation is just for motors in serial programming. There are no other time lags included, such as motor driver lag or microcontroller processing lag. Therefore, because of these time lags, the performance of these robotic technologies is facing issues, making them slower in bagging and harvesting.

2.3 Conclusion

This chapter delves into the intriguing world of robotics for apple bagging and the notable progressions achieved in this domain. The integration of computer vision, deep learning, and robotics has facilitated the realization of the objective of enhancing apple bagging practices through the utilization of intelligent robotic systems.

The significance of apple bagging as a safeguard against pests, diseases, and environmental factors was initially underscored. The incorporation of robotics in this process presents promising prospects for optimizing operations, minimizing human labor, and augmenting orchard management.

The chapter has explored the diverse innovative methodologies and strategies that researchers have devised for the automation of apple bagging. The efficacy of machine learning algorithms such as YOLO V5s and YOLO-V4 in precisely detecting and segmenting apples of varying colors, shapes, and sizes has been observed. Furthermore, the implementation of sophisticated techniques such as channel pruning, SLIC segmentation, and feature extraction modules has demonstrated noteworthy enhancements in detection precision, velocity, and resource optimization.

In addition, our research has delved into the integration of robotic systems with advanced algorithms that possess cognitive capabilities, enabling them to independently traverse orchards, execute manipulations with robotic arms, and proficiently package apples in a timely manner. The integration of computer vision, robotic arm manipulation, and algorithmic decision-making has the capacity to transform conventional apple bagging methodologies, enhancing their precision, cost-efficiency, and productivity.

In summary, the chapter's research underscores the significant progress achieved in the domain of robotics for apple bagging. The aforementioned developments not solely augment the output and effectiveness of apple cultivation, but also make a valuable contribution to sustainable agricultural methodologies by curtailing the need for manual labor, optimizing resource allocation, and minimizing wastage.

Prospective investigations and advancements in the field of apple bagging robotics are imperative for overcoming the obstacles encountered in practical orchard settings. The future of apple bagging robotics will be influenced by enhancements in detection precision, velocity, and resilience, as well as progressions in self-governing navigation and decision-making proficiencies.

In summary, the incorporation of robotics and artificial intelligence into the process of apple bagging presents a promising area of research. The synergy among researchers, engineers, and farmers plays a pivotal role in propelling innovative practices, promoting acceptance, and ultimately revolutionizing apple bagging techniques towards a more sustainable and efficient trajectory. The utilization of technology has the potential to transform the process of apple harvesting, thereby making a significant contribution to the progress of the agricultural industry.

References

1. Zhu Z et al (2018) Life cycle assessment of conventional and organic apple production systems in China. J Clean Prod 201:156–168. https://doi.org/10.1016/j.jclepro.2018.08.032
2. Wu Z, Pan C (2021) State analysis of apple industry in China. IOP Conf Ser Earth Environ Sci 831(1). https://doi.org/10.1088/1755-1315/831/1/012067
3. Liang X, Zhang R, Gleason ML, Sun G (2022) Sustainable apple disease management in China: Challenges and future directions for a transforming industry. Plant Dis 106(3):786–799. https://doi.org/10.1094/PDIS-06-21-1190-FE
4. Fruit FD, Inouye A, Ward M (2019) Report name: Fresh deciduous fruit annual report highlights 2022:1–43
5. De-An Z, Jidong L, Wei J, Ying Z, Yu C (2011) Design and control of an apple harvesting robot. Biosyst Eng 110(2):112–122. https://doi.org/10.1016/j.biosystemseng.2011.07.005

6. Gao H, Liu Y, Li D, Yu Y (2017) Vision localization algorithms for apple bagging robot. Proc. 29th Chinese Control Decis. Conf. CCDC 2017:135–140. https://doi.org/10.1109/CCDC.2017.7978080
7. Li Q, Jia W, Sun M, Hou S, Zheng Y (2021) A novel green apple segmentation algorithm based on ensemble U-Net under complex orchard environment. Comput Electron Agric 180(November 2020):105900. https://doi.org/10.1016/j.compag.2020.105900
8. Lv J, Wang F, Xu L, Ma Z, Yang B (2019) A segmentation method of bagged green apple image. Sci Hortic (Amsterdam) 246(November 2018):411–417. https://doi.org/10.1016/j.scienta.2018.11.030
9. Zhang C, Zou K, Pan Y (2020) A method of apple image segmentation based on color-texture fusion feature and machine learning. Agronomy 10(7). https://doi.org/10.3390/agronomy10070972
10. Fan P et al (2021) A method of segmenting apples based on gray-centered RGB color space. Remote Sens 13(6). https://doi.org/10.3390/rs13061211
11. Zou K, Ge L, Zhou H, Zhang C, Li W (2022) An apple image segmentation method based on a color index obtained by a genetic algorithm. Multimed Tools Appl 81(6):8139–8153. https://doi.org/10.1007/s11042-022-11905-4
12. Jia W, Zhang Z, Shao W, Ji Z, Hou S (2022) RS-Net: robust segmentation of green overlapped apples. Precis Agric 23(2):492–513. https://doi.org/10.1007/s11119-021-09846-3
13. Mirhaji H, Soleymani M, Asakereh A, Abdanan Mehdizadeh S (2021) Fruit detection and load estimation of an orange orchard using the YOLO models through simple approaches in different imaging and illumination conditions. Comput Electron Agric 191(June):106533. https://doi.org/10.1016/j.compag.2021.106533
14. Luo L, Tang Y, Lu Q, Chen X, Zhang P, Zou X (2018) A vision methodology for harvesting robot to detect cutting points on peduncles of double overlapping grape clusters in a vineyard. Comput Ind 99:130–139. https://doi.org/10.1016/j.compind.2018.03.017
15. Wang D, He D (2021) Channel pruned YOLO V5s-based deep learning approach for rapid and accurate apple fruitlet detection before fruit thinning. Biosyst Eng 210:271–281. https://doi.org/10.1016/j.biosystemseng.2021.08.015
16. Liu X, Zhao D, Jia W, Ji W, Sun Y (2019) A detection method for apple fruits based on color and shape features. IEEE Access 7:67923–67933. https://doi.org/10.1109/ACCESS.2019.2918313
17. Fan P et al (2021) Multi-feature patch-based segmentation technique in the gray-centered RGB color space for improved apple target recognition Agriculture 11(3). https://doi.org/10.3390/agriculture11030273
18. Ji W, Pan Y, Xu B, Wang J (2022) A real-time apple targets detection method for picking robot based on ShufflenetV2-YOLOX. Agriculture 12(6). https://doi.org/10.3390/agriculture12060856
19. Ji W, Gao X, Xu B, Pan Y, Zhang Z, Zhao D (2021) Apple target recognition method in complex environment based on improved YOLOv4. J Food Process Eng 44(11). https://doi.org/10.1111/jfpe.13866
20. Ge L et al (2022) Three dimensional apple tree organs classification and yield estimation algorithm based on multi-features fusion and support vector machine. Inf Process Agric 9(3):431–442. https://doi.org/10.1016/j.inpa.2021.04.011
21. Liu X, Jia W, Ruan C, Zhao D, Gu Y, Chen W (2018) The recognition of apple fruits in plastic bags based on block classification. Precis Agric 19(4):735–749. https://doi.org/10.1007/s11119-017-9553-2
22. Wang Y, Zhang Y, Pu Y, Zhang J, Wang F (2018) Design of a new fruit tree bagging machine. IOP Conf Ser Mater Sci Eng 452(4). https://doi.org/10.1088/1757-899X/452/4/042099
23. Zhang WJ, Zhang F, Zhang J, Zhang J (2021) Analysis of bagging trajectory of an intelligent mobile electrical robot in hail climate. J Phys Conf Ser 2033(1). https://doi.org/10.1088/1742-6596/2033/1/012047

24. Xia H, Zhen W, Chen D, Zeng W (2020) Rigid-flexible coupling contact action simulation study of the open mechanism on the ordinary multilayer fruit paper bag for fruit bagging. Comput Electron Agric 173(February 2019):105414. https://doi.org/10.1016/j.compag.2020.105414
25. Xia H, Zhen W, Chen D, Zeng W (2019) An ordinary multilayer fruit paper bag supplying device for fruit bagging. HortScience 54(9):1644–1649. https://doi.org/10.21273/HORTSCI14171-19
26. Hua Y, Yang B, Zhou XG, Zhao J, Li L (2016) A novel progressively delivered fruit bagging apparatus. J Appl Hortic 18(2):123–127. https://doi.org/10.37855/jah.2016.v18i02.21

Chapter 3
Apple's In-Field Grading and Sorting Technology: A Review

Jiangfan Yu, Zhao Zhang, Mustafa Mhamed, Dongdong Yuan, and Xufeng Wang

Abstract Quality inspection is crucial to raise people's living standards in the realm of processing fruits and vegetables. Since apples are one of the best-producing crops, post-harvest infield grading is advantageous for enhancing economic efficiency and lowering production costs. Apple infield grading and sorting equipment's technology have advanced quickly in recent decades. This study overviews the quality inspection developments in infield grading and sorting equipment. We present typical visual quality variables, including color, size, flaws, and internal quality inspection, as supplemental data for detection parameters. Later, Optical imaging methods such as visible light, near-infrared, hyperspectral/multispectral, and structured light are surveyed for apple quality assessment. Finally, the difficulties in commercializing the existing apple infield grading and sorting equipment are offered, including the need to get information about the whole area, uneven lighting, and high prices. Further provides an overview of the pertinent hardware and computational solutions. This chapter makes a comprehensive summary of apple infield grading and sorting technology. It points out the problems that must be faced in the follow-up development, which can guide future work.

Keywords Apple quality inspection · Apple infield grading and sorting · Optical imaging technology · Application

---

J. Yu · Z. Zhang (✉) · M. Mhamed
Key Laboratory of Smart Agriculture System Integration, Ministry of Education, Beijing 100083, China
e-mail: zhaozhangcau@cau.edu.cn

Key Laboratory of Agricultural Information Acquisition Technology, Ministry of Agriculture and Rural Affairs, China Agricultural University, Beijing 100083, China

College of Information and Electrical Engineering, China Agricultural University, Beijing 100083, China

D. Yuan
Sweet Fruit, Co., Ltd, Suqian 223839, Jiangsu, China

X. Wang
College of Mechanical and Electrical Engineering, Tarim University, Alar 843300, China
e-mail: wxf@taru.edu.cn

Z. Zhang and X. Wang (eds.), *Towards Unmanned Apple Orchard Production Cycle*, Smart Agriculture 6, https://doi.org/10.1007/978-981-99-6124-5_3

3.1 Introduction

Apples are cultivated extensively all over the globe and are the third most produced fruit in the world [1]. High-grade apples are becoming increasingly popular as customer demands for the look and quality of apples rise [2]. Though apples must be differentiated and priced according to quality using conventional post-harvest processing techniques, this impacts customers' willingness to buy and cannot optimize advantages. If apples are graded after harvest, different quality apples can be sold separately, increasing the economic added value of apples and increasing consumers' desire to buy, thereby maximizing sales benefits and improving the industry's core competitiveness [3]. Apple grading involves evaluating the apple fruit's size, colour, defects, and other features, plus classifying the fruit into different grades based on the assessment outcomes. According to grades, fresh apples are sold on the market, while processing facilities turn secondary apples into juices or preserves [4].

The 1980s saw the start of the research of automated equipment, which has since undergone commercialization. Large indoor grading equipment had recently been automated, and instances include the Compac Spectrim from Norwegian company Tomra, the MERLIN system from US company Industry Vision Automation, the AGGROBOT from French company Maf Roda, and the fruit sorting equipment from Chinese company REEMOON. These devices typically use machine vision, spectral detection, and other technologies. Indoor grading systems, however, are only appropriate for a small number of dispersed production sites and highly organized production regions. In addition, the large size of indoor grading equipment makes it inconvenient to transport, and high costs result in longer return cycles for farmers, which also limits the promotion and application of equipment [5, 6]. Apples of different qualities are picked in the field and stored together before being transported indoors for sorting. Studies have shown that during this process, the proportion of fruit that is discarded due to secondary losses and cross-contamination can reach (15–20%), causing significant economic losses to producers [7, 8]. Therefore, researchers have begun focusing on real-time infield grading technology for apples after harvest to compensate for manual and indoor equipment grading shortcomings and maintain the quality of apples. Unlike indoor industrial grading, infield grading separates fruit by grade and removes defective fruit before storage, improving efficiency and effectively reducing the probability of cross-contamination between fruits. Furthermore, using sizeable indoor equipment for grading, the storage, grading, and sorting costs paid by growers are similar to the selling price of apples [9]. Infield grading shortens the time that apples are in circulation, reducing costs in storage, transportation, and other links, and maximizing sales benefits [10]. Consequently, apple infield grading has more potential to increase economic added value and is one of the most influential technologies in the apple industry chain [9].

Currently, apple grading is mainly done manually or with automated equipment. The traditional manual method usually requires experienced personnel to inspect each fruit in detail. This method is time-consuming, subjective, labour-intensive, and can result in visual variations due to fatigue, making it difficult to adapt to large-scale orchard production [11, 12].

We will now look at the advancement of grading techniques, grading algorithm development mainly focuses on apple quality detection. The engineering fields of machine vision and computer vision, which integrate image processing, artificial intelligence, and visual tools, have advanced the intelligence of quality detection in agriculture [13–15]. Many scholars have also applied them to apple quality detection (i.e., apple grading). The evaluation of apple quality is divided into appearance quality (visible to the human eye) and internal quality [16]. Appearance quality includes size, colour, and defects [17]. Generally, after obtaining the surface information of the apple through imaging technology and spectral technology, feature extraction and analysis are performed using image processing, spectral analysis, and machine learning technologies. Internal quality includes hardness, sugar content, and soluble solid content. It is often assessed by looking at data on the transmission and reflection of invisible light. Nowadays, in-field apple grading mainly focuses on external factors due to cost and detection speed constraints. As a result, internal quality detection will be supplemental in this article (see Sect. 3.3), which will concentrate on the machine vision-based method for detecting exterior quality.

Considering it cannot satisfy the demands of small volume, cheap cost, high throughput, and remarkable resilience, infield apple grading is not used in commercial applications [4]. However, much research has been conducted on apple infield grading equipment and technology in the past few decades. Some reviews have also been published in related fields, such as reviews of fruit grading technology [18–20], machine vision in agricultural applications [21, 22], and non-destructive testing of fruit quality [23–27]. Here, our survey focuses on apple infield grading and sorting technology based on the following questions:

1. What are equipment technology's features, difficulties, and advancements across time?
2. Apple appearance inspection techniques, resources, colour modeling, size, and detection frameworks?
3. What application strategies were employed, and what significant outcomes did they yield?
4. What are the present problems and potential remedies?

Then, we'll provide an overview of the main problems, the techniques' effectiveness, and the difficulties they face; our future view will be briefly discussed. The remainder of this chapter is structured as follows: Sect. 3.2 and 3.3 introduce the appearance and internal quality detection methods of apple, focusing on the appearance quality, including color, size, and defect detection of apple. Section 3.4 discusses the challenges and possible solutions to infield grading and sorting of apples. Finally, Sect. 3.5 gives the review's overall conclusions.

3.2 Apple Appearance Quality Inspection

Apple sales are most significantly impacted by the quality of their outward look, which has three key components: colour, size, and defects. Customers subjectively relate an apple's exterior to its inherent attributes, influencing their willingness to purchase [2]. Research on detecting apple size and colour has been relatively mature, while the focus has shifted to detecting apple defects. Consequently, this section will describe studies on apple size and color before comprehensively summarizing current research on spotting apple flaws.

3.2.1 Apple Color

Apple colour can indicate maturity, and consumers prefer apples with higher colouration. Colour information is the most fundamental information of an image, and the colour images obtained by cameras are consistent with human vision [28]. Therefore, colour detection is usually performed using traditional machine vision systems. Currently, many colour spaces are used to represent object colours, including the RGB colour space, HIS (Hue, Intensity, and Saturation) colour space, L*a*b* (Lightness, a: the component from green to red, b: the component from blue to yellow) colour space, and YCbCr (Y: Luma, Cb: Chroma of blue, Cr: Chroma of red) colour space. The RGB colour space is the most commonly used in colour feature extraction [22].

In a fuzzy logic-based grading system designed by Lorestani et al. [29], the ratio of the R component to the sum of the R, G, and B channels was used as the colour feature of apples for colour grading. After the data was fuzzified and fuzzy rules were applied to the inputs of colour and size, the final grade was obtained through fuzzy inference and de-fuzzification using CoG (Centroid of Gravity). The results showed that the consistency of the system's grading results with those of human experts was 90.8%. Zou et al. [30] extracted 17 colour features (average colour gradient, variance, and colour coordinates) from each apple. They used genetic algorithms to combine the features into new feature parameters. They were integrated with the stepwise decision tree method for apple colour grading. The findings demonstrated that the SVMalgorithmrecognition accuracy was inferior to that of backpropagation artificial neural networks but superior to both. Zhang et al. [31] performed statistical analysis on the RGB values of apple images. They found that apples had more red and less blue colours, with R/B values usually above 1.4. Using the R/B value as the colour feature, the recognition rate was 77.9%. Still, the recognition rate was relatively low due to some backgrounds having similar colours to the apples, resulting in their R/B values greater than 1.4. Sofu et al. [32] avoided the influence of light that may penetrate the system by using the differences in red, green, and blue colours to extract the colour features. Three types of apples were tested in the experiment, and the grading rate ranged from 93.44% to 100% (Table 3.1).

Table 3.1 Summary of literature on the solutions for apple colour inspection

Author(s)	Colour space	Method(s)	Accuracy
Tao et al. [33]	HIS	LDA	90%
Lorestani et al. [29]	RGB	Fuzzy logic	90.8%
Xiaobo et al. [30]	RGB and HIS	OFP	–
Zhang et al. [31]	RGB	BP neural network	77.9%
Chauhan and Singh [34]	HIS	K-NN	95.12%
Sofu et al. [32]	RGB	C4.5 decision tree	>93.44%

Note LDA: linear discriminant analysis; K-NN: k-nearest neighbor; OFP: organization feature parameter; and BP: backpropagation

3.2.2 Apple Size

Apple size is currently the most commonly used indicator for apple grading, with most apple diameters ranging from 60 to 90 mm. Since they are an uneven sphere, apples are difficult to quantify using a single geometric parameter. Various techniques have been developed to extract parameters when estimating apple size, including taking the maximum value, average value, circumscribed circle, rectangle, and the approach of the maximum transverse section diameter.

Due to the significant difference in grey level between the apple edge and the background, there is a maximum grey level variation. Feng and Wang [35] extracted the apple centroid. They searched outward along the radius direction from the centroid until the full grey level variation point was found and marked as the edge point. The axial direction of the apple was determined by the radius sequence from the centroid to the edge point and symmetry. The maximum width perpendicular to the axial direction was considered the size of the apple. In two sets of experiments, the accuracy of axial detection reached 94.4%, and the absolute error of size detection was not more than 3 mm. Chen et al. [36] established a multiple linear regression model based on the maximum diameter of apples and the maximum transverse section diameter in three apple images to obtain the maximum transverse section diameter fitting function. The results showed that the accuracy up to 87.1%, which could meet the application requirements of online sorting. Mizushima and Lu [37, 38] proposed a comprehensive approach to estimate apples' equatorial diameter. This method found the fruit stalk calyx axis through the radius function and calculated the equatorial diameter according to this orientation. Their approach estimated the equatorial diameter of four varieties, and the results showed that the apples with errors within ± 5 mm reached 98.9%, and the "Jonagold" variety reached 100%. Sofu et al. [32] segmented the apple and calculated the height and width of the apple region in the four images. The maximum value of them was taken as the diameter of the apple. Wang et al. [39] proposed an apple size estimation method based on local point clouds. The geometric parameters of the apple shape were estimated using point cloud data, and the apple geometric model based on the ellipsoid surface equation was

Table 3.2 Summary of literature on the solutions for apple size inspection

Author(s)	Size parameter	Method(s)	Accuracy
Feng and Wang [35]	Max length	Determine the orientation	94.4%
Chen et al. [36]	Maximum cross-sectional diameter	Mechanical fixation	87.1%
Mizushima and Lu [37, 38]	Maximum equatorial diameter	Determine the orientation	98.9%
Sofu et al. [32]	Max length	Compute the width and height in four images	–
Wang et al. [39]	The smallest bounding rectangle	GA combines the point cloud	–
Zhao and Ai [40]	Max length	Edge detection	>99%

constructed. The genetic and particle swarm algorithms were used to determine the best matching model parameters and the projected apple size. The average diameter estimate error was under 2.35 mm compared to the actual value. The apple sorting machine designed by Zhao and Ai [40] detected the apple's size using machine vision technology. The edge of the apple was detected using the Laplacian operator, and the centre point of the apple was calculated to obtain the maximum diameter of the apple. The detection results of 500 apples were statistically analyzed, the error rate was controlled within 0.7%, and the accuracy met the requirements of the automatic sorting device (Table 3.2).

3.2.3 Apple Defects

Apple defects include bruises, scratches, scabs, cracks, rot, insect damage, etc. These defects can be visually perceived as different from normal skin tissue and can be detected through image changes in texture, intensity, and edge gradients. The ability to recognize the stem and calyx (S/C) region is a problem for machine vision of apple detection systems. The S/C region in the picture often has an intensity level comparable to the fault, making it easy to mistake it for the defect area and resulting in incorrect apple grades. Researchers have proposed effective solutions, including a mechanical method that fixes the orientation of the apple using an automatic device to determine the position of the S/C area. The procedure is time-consuming and can lead to incomplete apple inspection, making it unsuitable for field grading systems [41, 42]. Therefore, in practical field environments, obtaining full surface information about apples while maintaining the orientation of the fruit stably and quickly is a problem. One feasible solution proposed by scholars is to directly detect the S/C area at the algorithm level using methods such as image processing, deep learning, spectral techniques rather than orientate the apples by mechanic.

Based on the extensive literature review, defect detection includes single defect detection and S/C detection. The commonly used methods for detection are imaging and spectral techniques. Imaging techniques can obtain apples' image and spatial information, thereby obtaining their contour, surface texture, and other characteristics; spectral techniques can obtain apples' surface reflectance, internal physical structure, and chemical composition. In addition to traditional machine vision technology based on visible light imaging, spectral imaging technology, which combines imaging and spectral techniques, has attracted increasing attention for the non-destructive detection of apple defects. Specific examples include infrared spectral imaging technology, multispectral imaging technology, hyperspectral imaging technology, and structured light imaging technology [43].

3.2.3.1 Visible Light

As a technology widely used in the field of fruit quality inspection, machine vision systems (Fig. 3.1) based on visible light imaging capture RGB images that are perceived by the human eye and have fast processing speed and low cost for processing information on apple defects [22]. For RGB images in apple, defect detection, colour, object shape, texture features, and other information in the image are all effective information [44].

Some visible defects on the apples' surface have a significant colour difference compared to normal skin tissue. The colour of the defect is usually dimmer, and based on this characteristic, local or global colour segmentation methods can be utilized to detect apple defects [45]. To improve the robustness of the defect detection algorithm, Blasco et al. [46] conducted offline training in advance to assign pixels in the area to pre-defined categories: background, primary colour, secondary colour, type 1 damage, type 2 damage, and S/C. A Bayesian discriminant model was created using the R, G, and B colour components to calculate the probability that the selected area belongs to the above categories. Then, a quadratic discriminant model was generated using different covariance matrices for each class to achieve accurate

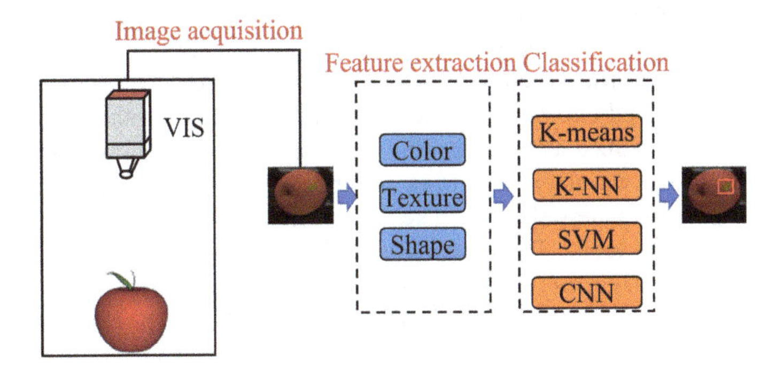

Fig. 3.1 Inspection of apple defects by visible light imaging

grading. Forty-eight apples were graded for flaws, and the detection findings were in line with the opinions of human specialists. In recognizing the S/C area, only five out of 100 apples were not identified.

Additionally, they carried out repeatability trials on the system, and the results showed that online fault detection had a repetition rate of roughly 86%, suggesting the system's stability. Moallem et al. [47] proposed a computer vision-based grading method for Golden Delicious apples, which uses pixel R, G, B, and H values as inputs to segment the S/C area and defects using an MLP neural network. The apple stem was divided into external visible and internal visible. The morphology was used to detect the externally visible stem due to the narrow width of the stem. The visible internal stem was detected using a Mahalanobis distance classifier by extracting the mean and covariance matrix of the stem pixel. The image was then converted to the YCbCr colour space, and the Cb component was used for K-means (K = 2) clustering to detect the calyx area. Based on the above results, the image was subtracted to remove the calyx area, thus retaining the true defects. When the stem is visible outside the apple, the approach has an excellent detection effect, but the accuracy is only 81% when the stem is within the apple. The algorithm achieved an accuracy rate of over 94% for calyx detection.

The texture feature is a similar global property to colour. It can perform well in apple flaw detection since it represents the arrangement features of the object surface with gradual or periodic change and has scale invariance. The commonly used texture analysis methods include statistical, structural, and signal processing methods. The statistical texture analysis method performs exceptionally well when the texture primitives are small (micro-texture). Arlimatti [48] proposed a window-based apple defect detection method based on statistical features. Firstly, the V channel in the HSV image is extracted and divided into several windows and the statistical features. After that, implemented KNN classifier for defect classification, and the S/C was regarded as a defect. The accuracy of apple defect detection was 96%, and the accuracy of S/C detection was 80%.

The Gabor transform is a signal processing technique for obtaining texture features that effectively get the target's local spatial and frequency domain data. It is very sensitive to the edge of the image and has strong adaptability to illumination changes. Therefore, the Gabor wavelet transform is widely used in visual information understanding of agricultural product detection. Jolly and Raman [49] proposed a method for detecting apple surface defects based on Gabor features. After preprocessing, the image was divided into several regions by the K-means algorithm (K = 3), and each region was assigned a corresponding gray value. After the grayscale image is obtained, the Gabor wavelet is used to extract the image features. Then the Haralick feature, local binary pattern (LBP) and kernel principal component analysis are used to obtain the feature vector. Finally, SVM and K-NN were employed to classify the samples into two categories (normal and defective) and four (normal, spot, scab and decay). The results show that the combination of Gabor + LBP can achieve more than 95% for the 4 classes. In the grading system designed by Vakilian and Massah [50], golden and red delicious apples were selected as samples. The Gabor filter is used to extract the image's energy, mean and variance features, and the apple is classified as the input layer node of the three-layer artificial neural network. Golden delicious and red delicious have accuracy rates of 89% and 92%, respectively, indicating the efficiency of the suggested strategy.

Texture feature extraction based on the model method considers that texture is formed based on a parameter-controlled distribution model, and the classification model is often applied to image analysis. Fractal is a term that describes the appearance of objects. Fractal theory indicates that most natural object surfaces are fractal in space [51]. Li et al. [52] proposed a method for identifying the S/C based on fractal dimension. Fractal features are scale-invariant, which describes the degree of surface roughness or irregular boundaries, and are unaffected by ambient light intensity and apple orientation changes. Based on this feature, a traditional fractal dimension and four directional fractal dimensions of the image are selected. Five fractal dimensions are used as features to input the back propagation network to classify the S/C and defect area. The accuracy rate is higher than 93%.

The shape is an intuitive feature, and when it comes to apples, people easily associate it with its external contour, namely the apple shape. Similarly, the shape exhibited by apple surface defects also has its unique characteristics, and many scholars have chosen to extract apple defects based on shape information. Eccentricity is a parameter that describes the shape of a region, and the lower the value, the closer the region is to a circle. In distinguishing apple defects (scab and rot) from S/C, Sofu et al. [32] used the K-means algorithm to cluster the image to detect a defect and S/C regions and calculated the eccentricity of the areas. The defect region's eccentricity is higher than the S/C region close to a circle. By combining color and eccentricity, the algorithm can distinguish the defect area from the S/C area and extract it.

Nevertheless, relying solely on the information provided by the image cannot accurately identify some early bruises and defects with low contrast or even invisible defects, which primarily exist under the skin and present a similar appearance to normal fruit tissue. This impacts the grading system's quality inspection [53, 54]. Many academics have suggested novel solutions to these problems, such as applying deep learning to replace conventional detection and identification techniques [55–57], or choosing other imaging methods to obtain more information about apple defects, including near-infrared, hyperspectral/multispectral imaging, and structured light (see next sections).

3.2.3.2 Near-Infrared

Infrared light refers to the band of wavelengths beyond visible light, which is divided explicitly into near-infrared (0.75–3 μm), mid-infrared (3–6 μm), and far-infrared (6–15 μm) [58]. Since Norris first applied near-infrared technology to agriculture, its non-contact and non-destructive detection mode has also been frequently used to analyze fruit quality parameters [59]. Since the emitting light will induce the vibration of most particular portions of molecules in fruits, the near-infrared spectral area may absorb pertinent information on organic compounds and the composition of molecular structures. So, appropriate models may be developed using chemometrics to study fruit quality. Near-infrared has become one of the most mainstream methods for non-destructive analysis of fruit quality [60]. Near-infrared-based apple defect detection methods can be divided into imaging analysis and spectral analysis.

Image analysis is similar to visible light analysis, which extracts relevant features such as shape through image processing to detect defects. The thickness ratio is a dimensionless shape parameter, and the closer its value is to one, the closer the area is to circular. Rehkugler and Throop [61] designed a grading system based on the near-infrared image to detect defects using the thickness ratio parameter of the extracted area. First, each row pixel of the original image is subjected to low-pass filtering, and each pixel value is replaced by the average value of 31 consecutive pixels starting from it. The newly generated image matrix is subtracted from the original grayscale image to eliminate slowly changing areas. The grading results show that the error grading rate of each level is approximately 10%. The defect objects detected by the above algorithm are bruises placed at room temperature for 24 h and the near-infrared reflectance of the bruised tissue changes with temperature and time. Therefore, an algorithm based on near-infrared reflectance to separate bruises was developed to distinguish between new and old bruises [62]. After low-pass filtering and morphological processing, the shape of the new and old bruises changes due to different reflectance. Therefore, based on the shape factor and threshold segmentation, the algorithm can distinguish the healing tissue within one day and two months. In the near-infrared band, the reflectance coefficient of the apple defect is lower than that of normal tissue. Liu and Wang [63] combined this feature with the apple's shape features to establish pixel-based defect detection rules for the apple grayscale image. The stem calyx and defect are distinguished based on area, centroid, and other shape features. Experiments with different threshold settings show that the detection rate of the defect area is above 90%, indicating that the system is not sensitive to threshold changes and has strong robustness. To further classify apple defects [64], for defect grading, they employed the BP neural network after classifying the defect types into bruising, punctures, cracking, insect damage, illness, and bug fruits. They also retrieved the defect region's width, aspect ratio, circularity, and other shape properties. The grading accuracy is above 50%. In the S/C detection system established by Zhang et al. [65], a feature construction method based on evolutionary construction (ECO) was proposed [65]. In contrast to the previous studies, the approach generates ECO features by image manipulation and iterative transformation, automatically extracting high-quality features from apple photos that are difficult to see with the naked eye. Then, a robust classifier that can accurately identify the defect and S/C in the near-infrared picture with (94.00%) accuracy was created by combining the AdaBoost and perceptron associated with the chosen ECO characteristics. The method is adaptive to different apple varieties and does not require designing corresponding code for each apple variety, making it highly generalizable. Hu et al. [66] obtained apple shape information from a 3D infrared imaging system and converted the 3D mesh into a 2D feature map, providing a learning method for 3D objects for deep learning models. CNN applied for the apple defect recognition. The test results showed that the highest defect recognition rate reached 97.67%, which exceeded the best feature extraction performance using shape descriptors. In addition to using a single near-infrared imaging technology, researchers often combine other spectral bands to identify the S/C, thereby achieving more accurate defect detection. Wen and Tao [67] found that a mid-infrared camera with a spectral range of 3–5 μm can shield apple damage areas and remain sensitive to the S/C area. Based on this characteristic, a dual-wavelength imaging method was proposed, in which the mid-infrared camera identifies the S/C. The near-infrared camera identifies the S/C and defect, and the defect area is extracted by combining the two. The article does not mention that in this dual-camera system, the cameras are usually installed in adjacent positions, which may cause differences in the obtained images. Cheng et al. [68] image registration and dual-image fusion based on this technique addressed the issue of picture discrepancies and retrieved the faulty region. Over 92.00% of defects are accurately detected online.

Spectral analysis detects defects by establishing quantitative and qualitative spectra models and associating them with relevant parameters [69]. Using a spectrophotometer, Xing et al. [70] obtained diffuse reflectance spectra in the visible and near-infrared ranges. The spectra correlation analysis was used to identify the effective bands. Then, a quadratic discriminant model with a total error of 16.3% for apples damaged for one day was used to classify bruised apples using Bayesian decision criteria. Therefore, based on previous research, Xing and de Baerdemaeker [71] improved the method by using a softening index related to the elastic modulus of apple tissue predicted by partial least squares to detect apple bruises, with an accuracy of more than (95%). The accuracy rate for fresh bruises (occurring within 1 h) was over 95%. Zhang et al. [72] utilized an integrating sphere system and an IAD algorithm to obtain the absorption and reduced scattering coefficient of early bruises in the apples' 400–1050 nm range. They successfully categorized the degree of bruising with a 92.7% accuracy rate by combining principal component analysis and support vector machines.

3.2.3.3 Hyperspectral/Multispectral Imaging

Hyperspectral imaging is a three-dimensional image data block composed of a series of continuous narrow-band images within a specific wavelength range (two-dimensional spatial information and one spectral dimension), whose data structure is usually referred to as a hypercube or data cube. The image consists of very narrow bands with high spectral resolution and can have several hundred or even thousands of bands. Multispectral imaging is similar to hyperspectral imaging and is considered an innovation of hyperspectral imaging [73]. Compared to the massive data volume and numerous spectral bands obtained by hyperspectral imaging, the multispectral imaging process is much simplified. Specifically, the spectral bands obtained by multispectral imaging systems are usually between 3 and 20, with wider bands and lower spectral resolution [22]. In image capture, these spectral bands are distinct and discontinuous. Therefore, multispectral photographs provide a group of isolated data points for each pixel instead of hyperspectral imaging, which produces a whole spectrum for each pixel [74, 75]. The imaging system significantly reduces

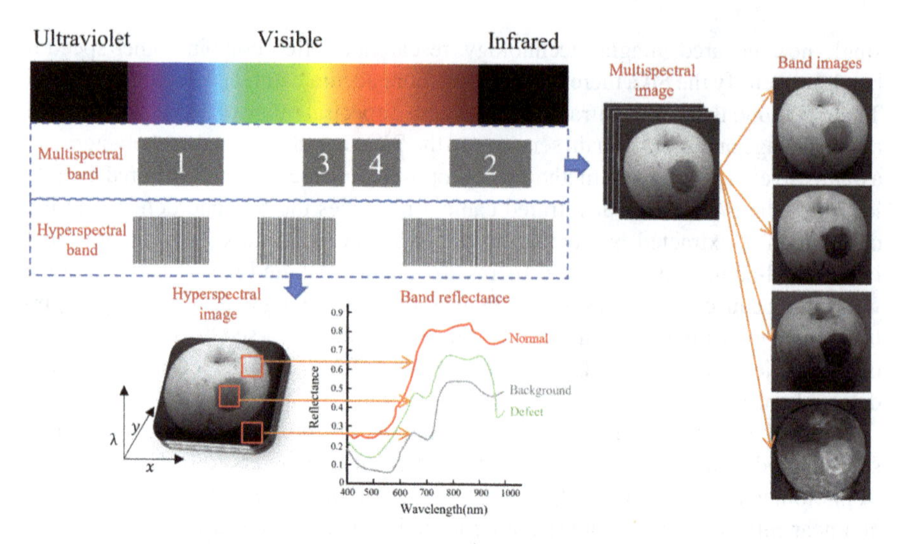

Fig. 3.2 The scheme of hyperspectral/multispectral imaging for apple defect detection

data volume, computational burden, and instrument costs. Figure 3.2 illustrates the variations between the two imaging techniques. Hyperspectral/multispectral imaging can provide sufficient external information about apples by combining spectra and images [76]. Still, a large number of spectral bands and data volume can have an impact on data processing. Therefore, statistical methods are generally used in data processing to reduce dimensions, remove redundant information, and identify feature wavelengths to improve detection speed [77, 78].

The steps for using hyperspectral/multispectral imaging to detect apple defects include data preprocessing, feature wavelength extraction, and defect detection. Data preprocessing mainly involves calibration and denoising of the image. Feature wavelength extraction removes redundant and duplicate bands and reduces data computation. Standard methods include principal component analysis (PCA), minimum noise fraction (MNF) transformation, genetic algorithms (GA), second derivative, stepwise analysis, and other methods. Defect detection uses some multivariate calibration models or image processing methods, such as K-means, partial least squares regression-discriminant analysis (PLS-DA), multiple linear regression (MLR), support vector machine (SVM), artificial neural network (ANN), Otsu's method, watershed segmentation, and others [79— 84].

Zhao et al. [85] collected hyperspectral image data in the 500–900 nm range to detect slight apple damage. Due to the uneven distribution of light intensity and dark current noise in each band, the image was calibrated using the calibration formula [86]. Then, data reduction was performed using principal component analysis, and the top four feature wavelengths were selected. By comparing the first four principal component images, it was found that 547 nm was the optimal wavelength for detecting slight apple damage. The image was subjected to a second derivative process centered on the 547 nm wavelength to eliminate bright spots, resulting in the final spectral image. Filtering, threshold segmentation, and erosion-dilation operations were performed on this image to detect apple defects, with a detection rate of approximately 89% for 60 apples. Zhang et al. [87] used the same method to identify damage and early rot in apples. However, since it was difficult to identify such defects using only a single wavelength image, six bands were selected as the feature wavelengths based on the band weight coefficients. The effectiveness of the feature wavelengths was verified by comparing the PCA transformation results based on these wavelengths and the full band.

Finally, the Otsu algorithm was used to detect defects, with an accuracy of 95.8% quickly. To distinguish the time of apple damage, Zhu and Li [88] proposed a defect detection method that can identify five time periods (1 min, 1 day, 2 days, 3 days, and 4 days later) of apple damage. Hyperspectral images in the 400–1000 nm range were collected, and the standard usual variable method was used to preprocess the data for smoothing and denoising. The minimum redundancy maximum relevance (mRMR) method was used to select feature wavelengths, and a grading model was established based on an extreme learning machine, partial least squares linear discriminant analysis, and a grading regression tree. The results demonstrated that the extreme learning machine, with a success rate of up to 96%, had the most incredible capacity to recognize damaged apples. By comparing the accuracy of different periods, it was found that the grading accuracy gradually increased as the damage time increased. Some scholars are interested in simultaneously detecting apple defects of different colours, believing this is significant for practical applications [89]. Therefore, a defect detection algorithm based on multispectral imaging combined with watershed segmentation was proposed to identify early bruises of apples with different peel colours. After obtaining multispectral images of three types of peel colours (green, intermediate, and red), the images were preprocessed using flat field correction and average normalization correction, and a band that could be adapted to three colours was selected using principal component analysis. An improved method was proposed to detect defects in sample images based on traditional watershed segmentation to address the over-segmentation problem. The identification accuracy reached 99.5%. In addition, hyperspectral imaging was also used to help distinguish S/C and defects of apples [90]. First, six feature wavelengths were obtained using principal component analysis, which was used to determine the PC1 image as a representative of the apple contour. The normal apple surface is smooth, close to a spherical shape, and the damaged area will have a particular depression.

In contrast, the depression at the S/C is more remarkable, indicating that the surface smoothness is worse. This characteristic can be reflected in the contour map of the apple's surface. For a stable surface, the lines between lines are approximately parallel, and the contour lines in the defect area will be slightly deformed. In contrast, the contour map of the S/C area is very distorted. Therefore, defects and S/C can be distinguished by segmenting the PC1 image and using the above principle. 160 Jonagold apples were tested, and the grading accuracy of the S/C was 98.3%. One key reason for misgrading was that the S/C area near the edge of the image caused a part of the surrounding area to be disconnected, affecting the algorithm detection.

3.2.3.4 Structured Illumination

Structured light technology projects a unique pattern onto the surface of an apple in three-dimensional space and observes the distortion of the surface imaging through another camera, thus detecting apple defects based on the deformation information (see Fig. 3.3). Generally, structured light can be divided into point structured light, line scanning structured light, and area array structured light [91, 92]. Using point-structured light, a laser projects a beam onto an object's surface. A better alternative to it light is line-scanning structured light, which reduces the complexity of point-structured light by transforming the scanning points into lines. Area array structured light projects an encoded pattern onto the surface of an object and obtains information after modulation. The latter two are more commonly used in practical applications among these projection methods.

In diffuse illumination scenes, defects, stem, and calyx areas appear as dark spots in the image. Crowe and Delwiche [95] extracted the illumination lines in structured illumination scenes and calculated their second derivatives to identify the position of the S/C area and remove it. The approach had a false positive rate of 25% for good apples and a false positive favorable rate of over 30% for bruises, cracks, and cuts. Yang [94] further utilized a flooding algorithm to separate dark spot regions from diffuse light photos based on the variations in how images look in two scenarios. The apple's three-dimensional surface was constructed based on structured light images, and stripe patterns were identified based on curvature analysis. The feature information of the two images was extracted and fed into a feedforward neural network to fuse two-dimensional and three-dimensional information for grading dark spot areas, with an average accuracy of (95%).

Lu and Lu [96] Employed dual-frequency and triple-frequency sine patterns with structured illumination reflection imaging to find bruising on the surface of apples.

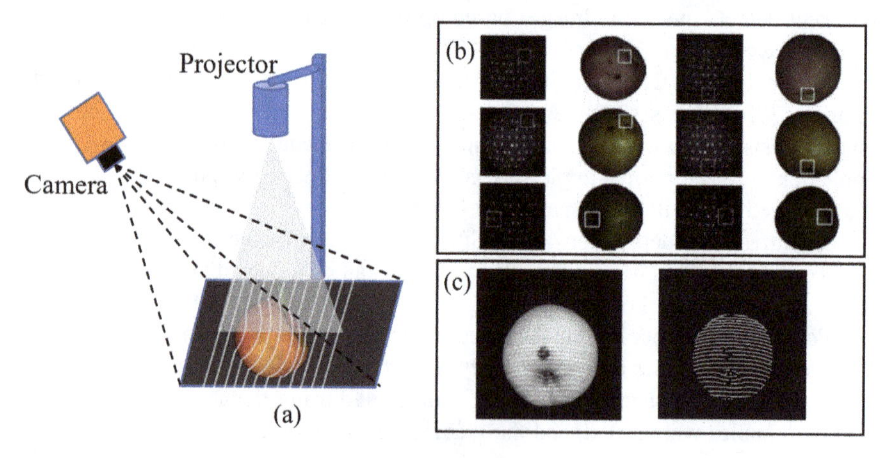

Fig. 3.3 **a** The scheme of structured illumination imaging, **b** near-infrared-coded dot matrix structured light patterns [93], **c** stripe projection [94]

With apple bruising apparent in photos at various frequencies, they developed three Fourier transform-based approaches to recover amplitude images at various frequencies. The method can obtain information in the spatial frequency domain and form four-dimensional enhanced data compared to spectral imaging mode. According to the findings of the experiments, a dual-frequency mode and "phase shift SPT" or "phase shift method" might produce a significant visual contrast. Still, the accuracy of defect recognition was not reported. As the concave portion of the apple S/C can cause deformation of pattern stripes, structured light imaging is commonly used to distinguish the apple S/C to improve the accuracy of defect recognition.

Zhang et al. [93] used a near-infrared structured light system to project near-infrared-coded dot matrix structured light patterns onto the surface of apples and then captured the coded image. The S/C could be identified by analyzing the relationship between the dot matrix and the projected pattern. Experiments were conducted on three varieties of apples, with an overall average recognition rate of (90.20%).

3.2.3.5 Other Method

Various practical imaging technologies in addition to prior approaches are utilized for apple flaw identification. Magnetic resonance imaging can obtain high-quality 2D and 3D images and has been used in research on food quality evaluation in recent years. Zion et al. [97] proposed a method that combines magnetic resonance imaging to detect apple damage. The technique initially determines the apple's center and radius, subsequently employing these values to subtract the apple core region from the picture and only keep the outside circular flesh area. Damaged tissue and apple core areas seem brighter in the image because they are in the damaged tissue. The picture is then segmented, and the damage is extracted using a suitable threshold based on pixel intensities. This method can fully distinguish bruised apples, but when acquiring MR images, the bruised apple area is manually set as the imaging plane, and it does not have the performance of online automatic detection. The idea that any item with a temperature over absolute zero will generate thermal radiation (radiant electromagnetic waves) and that size is proportional to temperature underlies the non-contact, quick, and highly accurate measuring method known as thermal imaging. Therefore, the thermal distribution field on the surface of an apple can be obtained using thermal imaging technology, and defect information can be obtained by obtaining a thermal image [58]. Baranowski et al. [98] believed that the thermal characteristics of fruit tissue with internal defects change, and defects could be identified using the different thermal characteristics on the surface of the fruit. Based on this characteristic, a method for early detection of apple damage based on thermal imaging technology has been proposed. After obtaining the thermal image, the image brightness and contrast are optimized by histogram analysis, noise reduction is achieved using a median filter, and defect regions are extracted based on image thresholding and recursive segmentation. The method can effectively detect defects and can distinguish bruises of different depths in the thermal spectrum sequence, verifying the linear correlation between bruise depth and thermal response frequency. When light shines on an apple, some photons are reflected and carry physical, colour, and other information. Laser backscattering imaging can achieve apple-quality inspection based on the absorption and scattering of this light [99]. Wu et al. [43] processed the laser backscattering spectral image of apples, removed noise using a Gaussian filter, and then extracted the laser-illuminated part as a training sample. A CNN model was applied to identify the defect area, average area, and S/C area with an accuracy of over 90.00%.

3.3 Internal Quality Inspection

Over time, people's concern for fruit quality has expanded beyond outward characteristics, including interior characteristics like sugar concentration, hardness, and soluble solids content [100, 101]. Delwiche and Sarig [102] developed an algorithm for analyzing the shape characteristics of impact force in an automatic apple hardness detection system. The results showed that the apple's surface was locally damaged during the impact process, and the large local variation in apple surface hardness resulted in a low correlation between sensor data and hardness measurement. Lammertyn et al. [103] used near-infrared spectroscopy combined with multivariate calibration techniques to establish a relationship model between reflectance spectra and acidity, soluble solids content, and the hardness of apples. The model's correlation coefficient was more significant than 0.8, indicating that near-infrared technology can be used for non-destructive detection of the internal quality of apples. Steinmetz et al. [104] combined two non-destructive sensors to predict the sugar content of apples using spectrophotometry and image processing. They found that colour information improved apple sugar content prediction outcomes, with an accuracy of 78% and a time of 3.5 s per apple. [105] quantified the backscattering profile of apples in five spectral bands ranging from 670 to 1060 nm. They used the scattering rate of different spectral bands as inputs to a neural network to predict red delicious apples' hardness and sugar content. In estimating apples' hardness and sugar content, the findings revealed that the multispectral combination of 880 nm, 905 nm, and 940 nm worked well. Liu et al. (2003) used intelligent fiber optic sensors to study the non-destructive detection of Fuji apple sugar content using three multivariate calibration algorithms (principal component regression, partial least squares, and stepwise regression) to process and analyze spectral data. They established the viability of fiber optic sensing methods for measuring apple sugar content. The potential of near-infrared Fourier transform techniques for fiber optic sensors is being further assessed. Yin et al. (2004) used partial least squares analysis to establish the relationship between diffuse reflectance spectra and apple sugar content. First and second derivative operations were used to analyze the spectral data, and results indicated that the second derivative spectral data had the most vital predictive power, allowing FTNIR approaches to estimate the amount of sugar in apples precisely. Ren et al. [106] used terahertz (0.75–1.1 THz) to monitor apple slices' water content. They applied Fourier transform, inverse transform, wavelet decomposition, and frequency domain feature extraction to the observed data. SVM, K-NN, and D-Tree were used to calculate the water content precisely. According to the experimental findings, over four consecutive days, all three classifiers managed to attain an accuracy of 95%. Apple internal quality detection is seldom carried out by infield grading equipment since it is highly dependent on the precision and stability of the spectral response.

3.4 Current Challenges and Possible Solutions

3.4.1 Whole Surface Information

In an ideal appearance quality inspection, acquiring complete surface information about the apple is necessary without repetitive testing on the same surface. Missing surface information can lead to the omission of critical details, resulting in the misclassification of apple grades, while redundant information can decrease the accuracy and efficiency of the inspection.

The current solutions include hardware and software components. On the hardware side: (1) Multiple images of the apple are obtained by rotating it to achieve full surface coverage. (2) Multiple cameras capture images of the apple from different angles. (3) Reflective images of the apple and camera blind spots are obtained by installing mirrors in the imaging chamber. By using conveyor belts or screw conveyors to rotate the apple, as long as a sufficient number of images are captured, complete surface information can be obtained. However, this approach leads to redundant information. For example, in calculating the colouration rate of the apple, overlapping regions in the images result in an inaccurate measurement of the actual colour area ratio. Methods involving multiple cameras or mirror installations may have undetectable areas due to the apple maintaining a fixed position. Additionally, reflective mirror images can have distortions and deformations, and multiple cameras can result in repetitive detection.

On the algorithm side, developing a 3D surface reconstruction algorithm for apple surfaces is a promising research direction. Zhu et al. [107] treated the apple as a Lambertian object, evaluated its surface reflectance map, and reconstructed the 3D surface information using the fast Fourier transform (FFT) with the shadow shape method. This method extended 2D near-infrared imaging to 3D reconstruction and significantly reduced the computation time through FFT. Zhang et al. [108] employed near-infrared linear array structured illumination and computer vision to encrypt height information for 3D reconstruction of the apple surface. Lu and Lu [109] proposed a phase analysis based on a structured illumination imaging system for reconstructing the 3D shape of the apple.

3.4.2 Uneven Illumination

Due to the apple's geometric approximation to a sphere, the surface brightness of spherical objects is uneven under illumination; This manifests in images as a bright or overexposed centre region with relatively darker boundaries. The uneven brightness distribution poses challenges for colour and defect detection in apples. The brighter center region exhibits colour distortion, resulting in significant color differences with the actual apple colour and affecting the classification of colours by grading systems. On the other hand, the border area is closer to the intensity of defect regions due to its relative darkness, which causes errors in defect categorization and lowers the overall system accuracy.

To address this issue, some researchers have proposed solutions by designing imaging chambers to mitigate the uneven brightness distribution. The design of imaging chambers revolves around diffusing the light, such as incorporating diffusers to alter the light propagation path [46, 56] or ensuring that the light reaches the camera at a 45-degree angle relative to the light source [110]. However, can prove inadequate owing to the complexity of field conditions, making brightness adjustment algorithms a feasible option.

Tao and Wen proposed an adaptive spherical transformation method [111] that converts the original spherical image into a planar image, addressing the issue of uneven intensity distribution caused by the apple's curved surface structure. This method effectively eliminates the influence of brightness variation on defect detection. Gómez-Sanchís et al. [112] considered the fruit surface a Lambertian surface and constructed a digital elevation model to obtain the fruit's geometric characteristics for correction. This method achieves uniform radiation intensity across the entire fruit surface but requires modeling for each fruit, making the process complex and increasing computational time. Zhang et al. [113] employed an automatic brightness correction method in the apple defect detection system. They identified the circular area, created circular masks to generate the apple's R-component circular picture, and then used an equation to adjust the image. Based on the fruit's size and form, this algorithm automatically modifies itself. However, the correction procedure may impact the detection precision when faults are found in the apple's core.

3.4.3 High Cost

A newly produced product must meet specific criteria, including being inexpensive, to be used in agriculture and replace human labour. Cost management is thus essential for apple field grading machinery. In addition to conventional visible light imaging, hyperspectral/multispectral cameras and thermal imaging can also capture apple images and perform well in quality inspection. However, they either have slow processing speeds, significant time costs, or high purchasing costs that make them difficult to apply in field settings. This issue is expected to be resolved with the reduction of hardware costs and the improvement of real-time capabilities in algorithms.

3.5 Conclusion and Future Outlook

In this chapter, we provide a comprehensive review of the developments in machinery and technology for Apple in-field grading:

1. We discussed numerous newly discovered optical detection approaches created for Apple quality inspection. We also thoroughly analyzed how various optical imaging techniques were utilized to identify apple defects.
2. We discussed the most recent applications and shortages and highlighted major issues in each section.
3. We offered the main challenges, such as acquiring whole surface information, uneven illumination on the spherical surface, high costs of spectral detection, and the robustness and generalization capability of algorithms. Futhermore, we provided the possible remedies, and our future views.

All that assists the researcher's communities in continuously developing the technology in apple sorting and grading systems, improving quality and production.

References

1. FAO (2022) Agricultural production statistics. 2000–2021. FAOSTAT Analytical Brief Series No. 60. Rome. https://doi.org/10.4060/cc3751en
2. Brosnan T, Sun D-W (2004) Improving quality inspection of food products by computer vision—a review. J Food Eng 61(1):3–16. https://doi.org/10.1016/s0260-8774(03)00183-3
3. Abbaspour-Gilandeh Y, Aghabara A, Davari M, Maja JM (2022) Feasibility of using computer vision and artificial intelligence techniques in detection of some apple pests and diseases. Appl Sci 12(2):906
4. Zhang Z, Lu Y, Lu R (2021) Development and evaluation of an apple infield grading and sorting system. Postharvest Biol Technol 180
5. Ismail N, Malik OA (2022) Real-time visual inspection system for grading fruits using computer vision and deep learning techniques. Inf Process Agric 9(1):24–37
6. Yang M, Kumar P, Bhola J, Shabaz M (2021) Development of image recognition software based on artificial intelligence algorithm for the efficient sorting of apple fruit. Int J Syst Assur Eng Manag 13:322–330
7. Lu Y, Lu R, Zhang Z (2022) Development and preliminary evaluation of a new apple harvest assist and in-field sorting machine. Appl Eng Agric 38(1):23–35. https://doi.org/10.13031/aea.14522
8. Zhang Z, Pothula AK, Lu R (2017) Development and preliminary evaluation of a new bin filler for apple harvesting and in-field sorting machine. Trans ASABE 60(6):1839–1849. https://doi.org/10.13031/trans.12488
9. Mizushima A, Lu R (2010) Cost benefits analysis of in-field presorting for the apple industry. Paper presented at the 2010 Pittsburgh, Pennsylvania, June 20–June 23, 2010

10. Zhang Z, Pothula AK, Lu R (2017) Economic evaluation of apple harvest and in-field sorting technology. Trans ASABE 60(5):1537
11. Lu R, Zhang Z, Pothula A (2017) Innovative technology for enhancing apple harvest and postharvest handling efficiency. Fruit Q 25(2):11–14
12. Zhang Z (2015) Design, test, and improvement of a low-cost apple harvest-assist unit
13. Bhargava A, Bansal A (2021) Fruits and vegetables quality evaluation using computer vision: a review. J King Saud Univ-Comput Inf Sci 33(3):243–257. https://doi.org/10.1016/j.jksuci.2018.06.002
14. Narendra V, Hareesha K (2010) Quality inspection and grading of agricultural and food products by computer vision-a review. Int J Comput Appl 2(1):43–65
15. Patel KK, Kar A, Jha S, Khan M (2012) Machine vision system: a tool for quality inspection of food and agricultural products. J Food Sci Technol 49:123–141
16. Huang XL, Zheng JQ, Zhao MC (2007) Review on fruit grading supporting technology. J Nanjig Forestry Univ (Natural Sciences Edition) 31:123–126
17. Li Q, Wang M (2000) Development of automatic apple grading hardware system based on computer vision. Trans Chin Soc Agric Mach 31(2):56–59
18. Behera SK, Rath AK, Mahapatra A, Sethy PK (2020) Identification, classification & grading of fruits using machine learning & computer intelligence: a review. J Ambient Intell Humized Comput: 1–11. https://doi.org/10.1007/s12652-020-01865-8
19. Kumar A, Gill GS (2015) Automatic fruit grading and classification system using computer vision: a review. In: 2015 Second International Conference on Advances in Computing and Communication Engineering. IEEE, May, pp 598–603
20. Pandey R, Naik S, Marfatia R (2013) Image processing and machine learning for automated fruit grading system: a technical review. Int J Comput Appl 81(16):29–39
21. Dhiman B, Kumar Y, Kumar M (2022) Fruit quality evaluation using machine learning techniques: review, motivation and future perspectives. Multimed Tools Appl 81(12):16255–16277
22. Zhang B, Huang W, Li J, Zhao C, Fan S, Wu J, Liu C (2014) Principles, developments and applications of computer vision for external quality inspection of fruits and vegetables: a review. Food Res Int 62:326–343
23. Nicolai BM, Beullens K, Bobelyn E, Peirs A, Saeys W, Theron KI, Lammertyn J (2007) Nondestructive measurement of fruit and vegetable quality by means of NIR spectroscopy: a review. Postharvest Biol Technol 46(2):99–118
24. Wang H, Peng J, Xie C, Bao Y, He Y (2015) Fruit quality evaluation using spectroscopy technology: a review. Sensors 15(5):11889–11927
25. Zhang B, Gu B, Tian G, Zhou J, Huang J, Xiong Y (2018) Challenges and solutions of optical-based nondestructive quality inspection for robotic fruit and vegetable grading systems: a technical review. Trends Food Sci Technol 81:213–231
26. Chandrasekaran I, Panigrahi SS, Ravikanth L, Singh CB (2019) Potential of near-infrared (NIR) spectroscopy and hyperspectral imaging for quality and safety assessment of fruits: an overview. Food Anal Methods 12:2438–2458
27. Lu R, Van Beers R, Saeys W, Li C, Cen H (2020) Measurement of optical properties of fruits and vegetables: a review. Postharvest Biol Technol 159:111003
28. Zheng C, Sun D-W, Zheng L (2006) Recent developments and applications of image features for food quality evaluation and inspection—a review. Trends Food Sci Technol 17(12):642–655
29. Lorestani A, Omid M, Bagheri-Shooraki S, Borghei A, Tabatabaeefar A (2006) Design and evaluation of a fuzzy logic based decision support system for grading of Golden Delicious apples. Int J Agric Biol 8(4):440–444
30. Xiaobo Z, Jiewen Z, Yanxiao L (2007) Apple color grading based on organization feature parameters. Pattern Recogn Lett 28(15):2046–2053
31. Zhang Y, Li M, Qiao J, Liu G (2008) A segmentation algorithm for apple fruit recognition using artificial neural network. Paper presented at the Proceedings 36th International Symposium 'Actual tasks on agricultural engineering', Opatija, Croatia, 11–15 veljače, 2008

32. Sofu M, Er O, Kayacan M, Cetişli B (2016) Design of an automatic apple sorting system using machine vision. Comput Electron Agric 127:395–405
33. Tao Y, Heinemann P, Varghese Z, Morrow C, Sommer Iii H (1995) Machine vision for color inspection of potatoes and apples. Trans ASAE 38(5):1555–1561
34. Chauhan APS, Singh AP (2012) Intelligent estimator for assessing apple fruit quality. Int J Comput Appl 60(5):35–41
35. Feng B, Wang M (2003) Detecting method of fruit size based on computer vision. Trans CSAM 43(1):73–75
36. Chen Y, Zhang J, Li W, Ren Y, Tan Y (2012) Grading method of apple by maximum cross-sectional diameter based on computer vision. Trans Chin Soc Agric Eng 28(2):284–288
37. Mizushima A, Lu R (2011). Development of a cost-effective machine vision system for infield sorting and grading of apples: fruit orientation and size estimation. Paper presented at the 2011 Louisville, KY, August 7–10, 2011
38. Mizushima A, Lu R (2013) A low-cost color vision system for automatic estimation of apple fruit orientation and maximum equatorial diameter. Trans ASABE 56(3):813–827
39. Wang H, Yan R, Zhou X (2019) Apple shape index estimation method based on local point cloud. Trans Chin Soc Agric Mach 50(5):205–213
40. Zhao D, Ai Y (2022) Research on apple size detection method based on computer vision. Agric Mech Res 44(07):206–209 + 214
41. Penman DW (2001) Determination of stem and calyx location on apples using automatic visual inspection. Comput Electron Agric 33(1):7–18
42. Unay D, Gosselin B (2007) Stem and calyx recognition on 'Jonagold'apples by pattern recognition. J Food Eng 78(2):597–605
43. Wu A, Zhu J, Ren T (2020) Detection of apple defect using laser-induced light backscattering imaging and convolutional neural network. Comput Electr Eng 81:106454
44. Leemans V, Magein H, Destain MF (2000) On-line fruit grading according to external quality using machine vision. Biosyst Eng 83:397–404
45. Leemans V, Magein H, Destain M-F (1998) Defects segmentation on 'Golden Delicious' apples by using colour machine vision. Comput Electron Agric 20(2):117–130
46. Blasco J, Aleixos N, Moltó E (2003) Machine vision system for automatic quality grading of fruit. Biosys Eng 85(4):415–423. https://doi.org/10.1016/s1537-5110(03)00088-6
47. Moallem P, Serajoddin A, Pourghassem H (2017) Computer vision-based apple grading for golden delicious apples based on surface features. Inf Process Agric 4(1):33–40
48. Arlimatti SR (2012) Window based method for automatic classification of apple fruit. Int J Eng Res Appl 2(4):1010–1013
49. Jolly P, Raman S (2016) Analyzing surface defects in apples using Gabor features. Paper presented at the 2016 12th International Conference on Signal-Image Technology & Internet-Based Systems (SITIS)
50. Vakilian KA, Massah J (2016) An apple grading system according to European fruit quality standards using Gabor filter and artificial neural networks. Sci Study Research Chem Chem Eng Biotechnol Food Ind 17(1):75
51. Mandelbrot BB, Mandelbrot BB (1982) The fractal geometry of nature, vol 1. WH Freeman, New York
52. Li Q, Wang M, Gu W (2002) Computer vision based system for apple surface defect detection. Comput Electron Agric 36(2–3):215–223
53. Lü Q, Tang M (2012) Detection of hidden bruise on kiwi fruit using hyperspectral imaging and parallelepiped classification. Procedia Environ Sci 12:1172–1179
54. Nicolaï BM, Defraeye T, De Ketelaere B, Herremans E, Hertog ML, Saeys W, Torricelli A, Vandendriessche T, Verboven P (2014) Nondestructive measurement of fruit and vegetable quality. Annu Rev Food Sci Technol 5:285–312
55. Fan S, Li J, Zhang Y, Tian X, Wang Q, He X, Zhang C, Huang W (2020) On line detection of defective apples using computer vision system combined with deep learning methods. J Food Eng 286:110102

56. Fan S, Liang X, Huang W, Jialong Zhang V, Pang Q, He X, Li L, Zhang C (2022) Real-time defects detection for apple sorting using NIR cameras with pruning-based YOLOV4 network. Comput Electron Agric 193:106715. https://doi.org/10.1016/j.compag.2022.106715
57. Wang Z, Jin L, Wang S, Xu H (2022) Apple stem/calyx real-time recognition using YOLO-v5 algorithm for fruit automatic loading system. Postharvest Biol Technol 185:111808
58. Vadivambal R, Jayas DS (2011) Applications of thermal imaging in agriculture and food industry—a review. Food Bioprocess Technol 4:186–199
59. Norris KH (1964) Design and development of a new moisture meter. Agric Eng 45(7):370–372
60. Sun T, Huang K, Xu H, Ying Y (2010) Research advances in nondestructive determination of internal quality in watermelon/melon: a review. J Food Eng 100(4):569–577
61. Rehkugler G, Throop J (1986) Apple sorting with machine vision. Trans ASAE 29(5):1388–1397
62. Throop J, Aneshansley D, Upchurch B (1995) An image processing algorithm to find new and old bruises. Appl Eng Agric 11(5):751–757
63. Liu H, Wang M (1998) Automatic detection of defects on apple with the computer image technology. Trans Chin Soc Agric Mach 29(4):81–86
64. Liu H, Wang M (2004) Method for classification of apple surface defect based on digital image processing. Nongye Gongcheng Xuebao (Trans Chin Soc Agric Eng), 20(6):138–140
65. Zhang D, Lillywhite KD, Lee D-J, Tippetts BJ (2013) Automated apple stem end and calyx detection using evolution-constructed features. J Food Eng 119(3):411–418
66. Hu Z, Tang J, Zhang P, Jiang J (2020) Deep learning for the identification of bruised apples by fusing 3D deep features for apple grading systems. Mech Syst Signal Process 145:106922
67. Wen JZ, Tao Y (1998) Dual-wavelength imaging for online identification of stem ends and calyxes. Paper presented at the Applications of Digital Image Processing XXI
68. Cheng X, Tao Y, Chen Y, Luo Y (2003) Nir/MIR dual–sensor machine vision system for online apple stem–end/calyx recognition. Trans ASAE 46(2):551
69. Zhang B, Dai D, Huang J, Zhou J, Gui Q, Dai F (2018) Influence of physical and biological variability and solution methods in fruit and vegetable quality nondestructive inspection by using imaging and near-infrared spectroscopy techniques: a review. Crit Rev Food Sci Nutr 58(12):2099–2118
70. Xing J, Bravo C, Moshou D, Ramon H, De Baerdemaeker J (2006) Bruise detection on 'Golden Delicious' apples by vis/NIR spectroscopy. Comput Electron Agric 52(1–2):11–20
71. Xing J, De Baerdemaeker J (2007) Fresh bruise detection by predicting softening index of apple tissue using VIS/NIR spectroscopy. Postharvest Biol Technol 45(2):176–183
72. Zhang S, Wu X, Zhang S, Cheng Q, Tan Z (2017) An effective method to inspect and classify the bruising degree of apples based on the optical properties. Postharvest Biol Technol 127:44–52
73. Su WH, Sun DW (2018) Multispectral imaging for plant food quality analysis and visualization. Compr Rev Food Sci Food Saf 17(1):220–239
74. Feng C-H, Makino Y, Oshita S, Martín JFG (2018) Hyperspectral imaging and multispectral imaging as the novel techniques for detecting defects in raw and processed meat products: current state-of-the-art research advances. Food Control 84:165–176
75. Qin J, Chao K, Kim MS, Lu R, Burks TF (2013) Hyperspectral and multispectral imaging for evaluating food safety and quality. J Food Eng 118(2):157–171
76. Li JB, Rao XQ, Ying YB (2011) Advance on application of hyperspectral imaging to nondestructive detection of agricultural products external quality. Spectrosc Spectr Anal 31(8):2021–2026
77. Fan S, Li J, Xia Y, Tian X, Guo Z, Huang W (2019) Long-term evaluation of soluble solids content of apples with biological variability by using near-infrared spectroscopy and calibration transfer method. Postharvest Biol Technol 151:79–87
78. Yang L, Guo J (2013) Applied research of agricultural product non-destructive detection using hyperpectral imaging technology. J Agric Pap 6:1–7
79. ElMasry G, Wang N, Vigneault C, Qiao J, ElSayed A (2008) Early detection of apple bruises on different background colors using hyperspectral imaging. LWT-Food Sci Technol 41(2):337–345

80. Huang W, Li J, Wang Q, Chen L (2015) Development of a multispectral imaging system for online detection of bruises on apples. J Food Eng 146:62–71
81. Lu R (2003) Detection of bruises on apples using near–infrared hyperspectral imaging. Trans ASAE 46(2):523
82. Mahanti NK, Pandiselvam R, Kothakota A, Chakraborty SK, Kumar M, Cozzolino D (2022) Emerging non-destructive imaging techniques for fruit damage detection: image processing and analysis. Trends Food Sci Technol 120:418–438
83. Mehl PM, Chen Y-R, Kim MS, Chan DE (2004) Development of hyperspectral imaging technique for the detection of apple surface defects and contaminations. J Food Eng 61(1):67–81
84. Zhang B, Fan S, Li J, Huang W, Zhao C, Qian M, Zheng L (2015) Detection of early rottenness on apples by using hyperspectral imaging combined with spectral analysis and image processing. Food Anal Methods 8:2075–2086
85. Zhao J, Liu J, Chen Q, Saritporn V (2008) Detecting subtle bruises on fruits with hyperspectral imaging. Trans CSAM 39(1):106–109
86. Polder G, van der Heijden GW, Keizer LP, Young IT (2003) Calibration and characterisation of imaging spectrographs. J near Infrared Spectrosc 11(3):193–210
87. Zhang B, Huang W, Li J, Zhao C, Liu C, Huang D (2013) Detection of bruises and early decay in apples using hyperspectral imaging and PCA. Infrared Laser Eng 42(5), e13952
88. Zhu X, Li G (2019) Rapid detection and visualization of slight bruise on apples using hyperspectral imaging. Int J Food Prop 22(1):1709–1719
89. Luo W, Zhang H, Liu X (2019) Hyperspectral/multispectral reflectance imaging combining with watershed segmentation algorithm for detection of early bruises on apples with different peel colors. Food Anal Methods 12:1218–1228
90. Xing J, De Baerdemaeker J (2005) Bruise detection on 'Jonagold' apples using hyperspectral imaging. Postharvest Biol Technol 37(2):152–162
91. Geng J (2011) Structured-light 3D surface imaging: a tutorial. Adv Opt Photonics 3(2):128–160
92. Zhang S (2018) High-speed 3D shape measurement with structured light methods: a review. Opt Lasers Eng 106:119–131
93. Zhang C, Zhao C, Huang W, Wang Q, Liu S, Li J, Guo Z (2017) Automatic detection of defective apples using NIR coded structured light and fast lightness correction. J Food Eng 203:69–82
94. Yang Q (1996) Apple stem and calyx identification with machine vision. J Agric Eng Res 63(3):229–236
95. Crowe T, Delwiche M (1996) Real-time defect detection in fruit—part II: an algorithm and performance of a prototype system. Trans ASAE 39(6):2309–2317
96. Lu Y, Lu R (2017) Using composite sinusoidal patterns in structured-illumination reflectance imaging (SIRI) for enhanced detection of apple bruise. J Food Eng 199:54–64
97. Zion B, Chen P, McCarthy MJ (1995) Detection of bruises in magnetic resonance images of apples. Comput Electron Agric 13(4):289–299
98. Baranowski P, Mazurek W, Witkowska-Walczak B, Sławiński C (2009) Detection of early apple bruises using pulsed-phase thermography. Postharvest Biol Technol 53(3):91–100
99. Pathmanaban P, Gnanavel B, Anandan SS (2019) Recent application of imaging techniques for fruit quality assessment. Trends Food Sci Technol 94:32–42
100. Guo Z, Huang W, Peng Y, Chen Q, Ouyang Q, Zhao J (2016) Color compensation and comparison of shortwave near infrared and long wave near infrared spectroscopy for determination of soluble solids content of 'Fuji' apple. Postharvest Biol Technol 115:81–90
101. Mendoza F, Lu R, Cen H (2014) Grading of apples based on firmness and soluble solids content using Vis/SWNIR spectroscopy and spectral scattering techniques. J Food Eng 125:59–68
102. Delwiche M, Sarig Y (1991) A probe impact sensor for fruit firmness measurement. Trans ASAE 34(1):187–0192
103. Lammertyn J, Nicolaï B, Ooms K, De Smedt V, De Baerdemaeker J (1998) Non-destructive measurement of acidity, soluble solids, and firmness of Jonagold apples using NIR-spectroscopy. Trans ASAE 41(4):1089

104. Steinmetz V, Roger J, Molto E, Blasco J (1999) On-line fusion of colour camera and spectrophotometer for sugar content prediction of apples. J Agric Eng Res 73(2):207–216
105. Lu R (2003) Predicting apple fruit firmness and sugar content using near-infrared scattering properties. Paper presented at the 2003 ASAE Annual Meeting.
106. Ren A, Zahid A, Zoha A, Shah SA, Imran MA, Alomainy A, Abbasi QH (2019) Machine learning driven approach towards the quality assessment of fresh fruits using non-invasive sensing. IEEE Sens J 20(4):2075–2083
107. Zhu B, Jiang L, Cheng X, Tao Y (2005) 3D surface reconstruction of apples from 2D NIR images. Paper presented at the Two-and Three-Dimensional Methods for Inspection and Metrology III
108. Zhang B, Huang W, Wang C, Gong L, Zhao C, Liu C, Huang D (2015) Computer vision recognition of stem and calyx in apples using near-infrared linear-array structured light and 3D reconstruction. Biosys Eng 139:25–34
109. Lu Y, Lu R (2017) Phase analysis for three-dimensional surface reconstruction of apples using structured-illumination reflectance imaging. Paper presented at the Sensing for Agriculture and Food Quality and Safety IX
110. Papadakis SE, Abdul-Malek S, Kamdem RE, Yam KL (2000) A versatile and inexpensive technique for measuring color of foods. Food Technol (Chicago) 54(12):48–51
111. Tao Y, Wen Z (1999) An adaptive spherical image transform for high-speed fruit defect detection. Trans ASAE 42(1):241
112. Gómez-Sanchís J, Moltó E, Camps-Valls G, Gómez-Chova L, Aleixos N, Blasco J (2008) Automatic correction of the effects of the light source on spherical objects. An application to the analysis of hyperspectral images of citrus fruits. J Food Eng 85(2):191–200
113. Zhang B, Huang W, Gong L, Li J, Zhao C, Liu C, Huang D (2015) Computer vision detection of defective apples using automatic lightness correction and weighted RVM classifier. J Food Eng 146:143–151
114. Yande L, Yibin Y, Huanyu J (2003) Experiments on measurement sugar content of fuji apple with an optical fiber sensor. Chin J Sens Actuators 16(03):328–331

Chapter 4
A Review of Apple Bagging Technology and Commercial Products on the Market

Yankun Ma, Kai Zhao, Zeheng Qian, and Afshin Azizi

Abstract Technology for bagging apples, which is a major initiative aimed at improving fruit quality and producing pollution-free fruit, plays an important role in the apple cultivation and production industry. In recent years, the development of automated apple bagging technology has been quite slow. Currently, there are two primary methods that offer the greatest feasibility for opening bags in this direction: pure mechanical bag opening technology and negative pressure suction cup bag opening technology. However, there is no more mature solution in bagging. Given the ongoing and significant development of automated apple bagging technology in the future, there remains a lack of comprehensive summary articles on this topic. Our chapter aims to fill this gap and contribute to the literature.

4.1 Introduction

During the period when young apple is still growing up, plenty of negative influence from natural environment around may cause damage to the young apple, such as pest, disease, and pesticide, which will lead to economic loss at the end of apple producing. To prevent pest and disease damage, reduce surface pollution and pesticide residue, apple bagging becomes a necessary option. After research to apples that have been bagged during their growth, scientists also discovered that bagging apple promotes surface coloring and improves fruit quality. Therefore, the value of apple bagging rises in apple producing field.

Y. Ma · K. Zhao · Z. Qian · A. Azizi (✉)
Key Laboratory of Smart Agriculture System Integration, Ministry of Education, China Agricultural University,
Beijing 100083, China
e-mail: azizi@cau.edu.cn

Key Laboratory of Agricultural Information Acquisition Technology, Ministry of Agriculture and Rural Affairs, China Agricultural University, Beijing 100083, China

College of Information and Electrical Engineering, China Agricultural University, Beijing 100083, China

Z. Zhang and X. Wang (eds.), *Towards Unmanned Apple Orchard Production Cycle*, Smart Agriculture 6, https://doi.org/10.1007/978-981-99-6124-5_4

4.1.1 Apple Young Fruit Image Pre-processing

After collecting high-resolution images of young apples, the first step is grayscale processing, which involves using the weighted average method. Each pixel in the collected high-resolution apple images is assigned a gray value using either the average or maximum method based on its RGB value. By traversing each pixel, a grayscale image is generated that reveals the characteristics and distribution of local and overall brightness, as well as chromaticity levels reflecting the entire image.

Recognizing young apple fruit in images can be challenging due to the similarity between their lime green color and that of the leaves in the background. Traditional color-based image recognition methods may not meet the requirements for reliable and accurate detection. To improve the reliability and accuracy of image recognition, the image should be binarized and processed using basic morphological operations such as erosion and expansion to refine the target area. It is worth noting that altering the order of these operations can yield different results. The process of applying morphological operations to the image involves two main steps: opening operation and closing operation. In the opening operation, erosion is followed by expansion, which smooths the edges of the image, removes small burrs, and breaks small joints [1]. In the closing operation, expansion is followed by erosion, which rounds the edges of the image, closes smaller gaps, and fills small holes. Both operations optimize image quality and facilitate later identification.

In cases where collected images are affected by external factors and contain a significant amount of noise, the smoothing filtering method can be used to reduce the noise and improve image quality. Two common denoising methods are the null domain method and the frequency domain method. Of these, the null domain method is generally preferred due to its faster computation time. The null domain method includes two filtering techniques: median filtering and domain average filtering. The former is effective in eliminating random noise, but can be sensitive to noise when calculating pixel values. On the other hand, the latter uses a mean filtering core, which may result in the loss of image details and create a blurring effect.

To identify young apple fruits and locate the target fruit, it is necessary to segment the image and extract the target region for subsequent analysis more accurately. The common method of threshold segmentation involves traversing the grayscale values of the pixels in an image and comparing them with a predetermined threshold value to perform image segmentation. The process of determining the threshold value for image segmentation involves two main methods: the optimal threshold method and the discriminant analysis method. Both methods involve calculating the probability distribution of pixel grayscale values. The former adjusts the gray level threshold to fit a normal distribution, while the latter computes the maximum variance of the gray levels to determine the appropriate threshold value [1] (Fig. 4.1).

Among the less common threshold segmentation methods, the Otsu method is an adaptive technique that divides an image into foreground and background regions

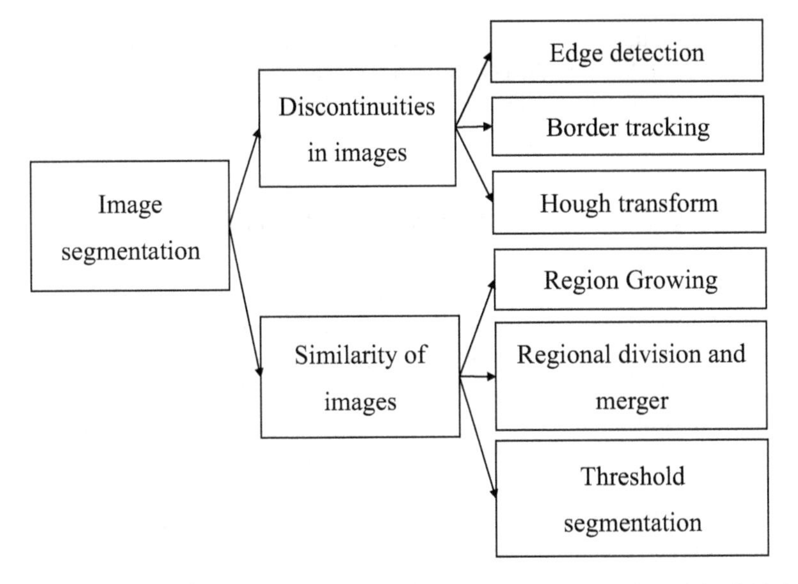

Fig. 4.1 Common methods of image segmentation. Image segmentation is mainly divided into discontinuities in images and similarity of images. Both two ways contain three common methods to process the image acquired

based on the grayscale characteristics of each pixel. It calculates the interclass variance between the regions and reduces the probability of misclassification by monitoring the magnitude of this variance during differentiation. The resulting threshold value is highly accurate and well-suited for performing segmentation operations on the processed images of young apple fruits.

4.1.2 Image Recognition

To recognize apple young fruit images, the image must first undergo binarization and connected component analysis using techniques such as connected domain labeling, labeling, and equivalent connected domain merging. After applying interference removal techniques such as thresholding to eliminate small areas, the image's feature terms can be identified. Among them, the contour edge extraction algorithm is mainly implemented by the 8-neighborhood connected chain code [1]. Since the young apple fruit is mainly round or oval in shape, the next step is to calculate and analyze the roundness of the extracted contour edges to make the most reasonable judgment.

To identify young apple fruits in the image, one can calculate the roundness of their contour edges and compare them with a preset roundness threshold. To accurately determine the position of young apple fruit and minimize interference from external forces such as wind, it is essential to measure the real-time swinging size of the fruit's center of gravity relative to the center of the screen. This enables precise location of the apple's center of gravity [1].

4.1.3 Visual Localization of Young Apple Fruits

The process of positioning young apple fruit relies on binocular stereo vision technology. The technology comprises of two types of stereo vision: convergent and parallel. While the former is simpler, its accuracy is more affected by distance. First, camera calibration is required to establish the position relationship between the camera lens and the target site, which consists of both linear and nonlinear models. Currently, the most used calibration method are Tsai's two-step method and Zhang plane calibration method. The convergent method combines linear and nonlinear calibration techniques to address the limitations of the linear model and streamline algorithm calculations. The parallel method optimizes the linear model with the nonlinear model by using images of the same target from different views. This approach builds on the two-step method used in the convergent method to enhance calibration accuracy. The fundamental aim of 3D reconstruction techniques using binocular stereo vision technology is to create a three-dimensional (3D) stereo representation of the scene using images captured by one or more cameras from different viewpoints. Traditional methods commonly use least squares to calculate 3D points for localization. However, in recent years, the use of OpenGL-based techniques for 3D localization has become more widespread. This approach involves depicting geometric shape units as endpoints and solving for each endpoint before rasterizing the image to obtain shape fragments [1]. This method is efficient and convenient, allowing for accurate 3D modeling based on flat images. As a result, other parameters such as the amount of wobbling of young apple fruit can also be measured.

To further enhance the localization capability, the image limit correction algorithm and the stereo matching algorithm can be optimized. The kernel of the image limit correction algorithm is to reduce the parallax between the two images from two dimensions to one dimension, which facilitates faster matching of the two images. The stereo matching algorithm, as the most critical stereo vision operation, is divided into local stereo matching and global stereo matching [1]. The former is faster but less accurate because of its simple calculation method; the latter is slower and more accurate, and the choice between the two needs to be considered according to the usage requirements (Fig. 4.2).

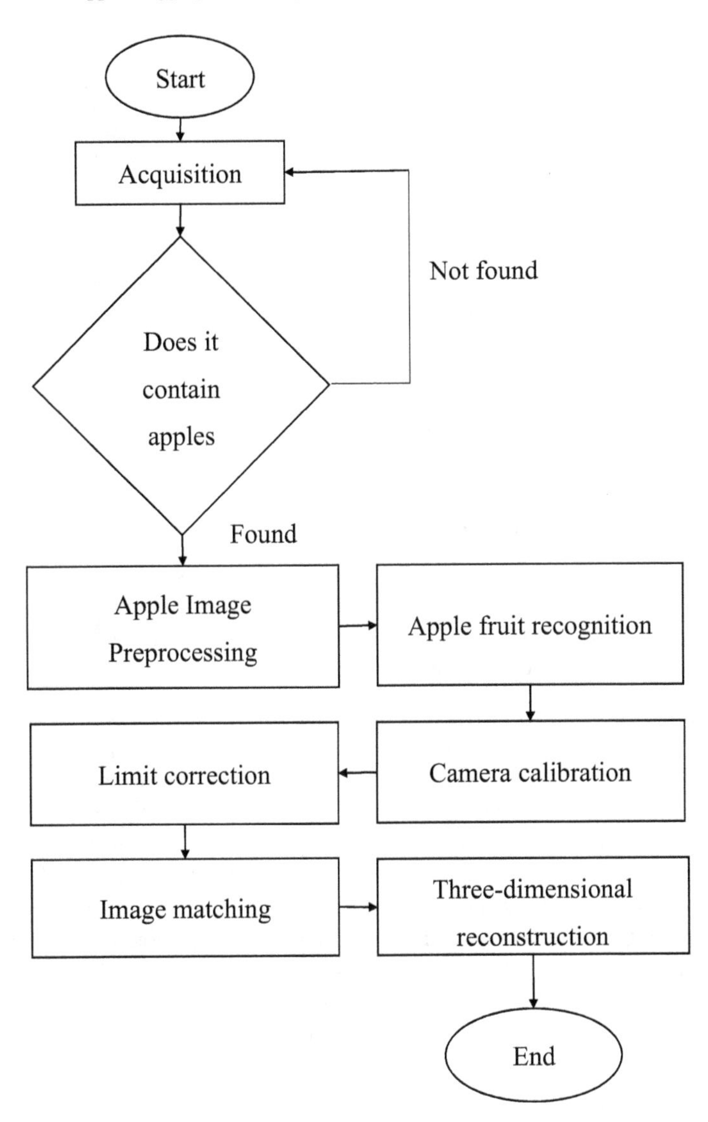

Fig. 4.2 The general process of the visual module. If apples are found during the acquisition, pre-set procedure of image processing will proceed. If not, acquisition will keep going

4.2 The Apple Fruit Bags Propped Open

Currently, bags are produced for a variety of uses, with the vast majority being either paper or plastic. Paper bags can be further classified as either single-layer or double-layered.

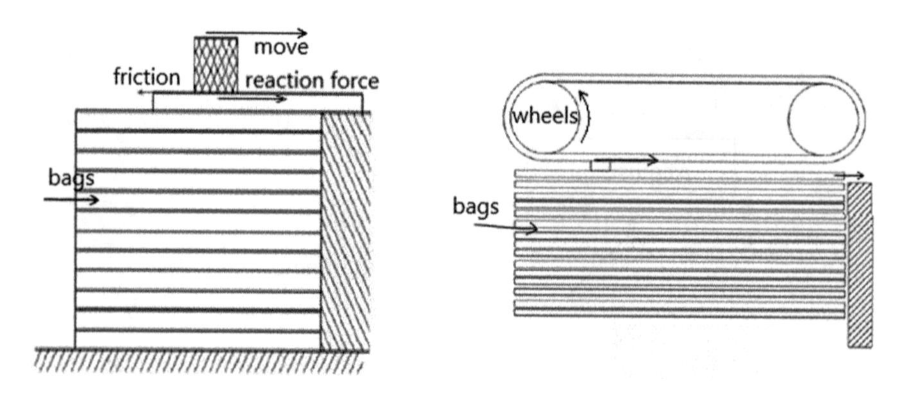

Fig. 4.3 Principle of using friction to separate paper bags. Both two graph use fiction to separate paper bags. Left one uses moving object to create fiction, while right one uses wheels

4.2.1 Separation of a Single Paper Bag

To improve efficiency, paper bags are typically stacked after being loaded into a bag storage device. To separate a single paper bag located at the top of the stack, there are various methods that can be used.

(1) Because paper bag surfaces are rough, separation can be accomplished through the friction created by rubbing against the surface of the bag [2]. To separate the top paper bag from a stack in the bag storage device, rough wheels can be added to the top of the device. When these wheels meet the surface of the paper bag, they create friction that causes the top bag to move away from the stack in a parallel motion, achieving the desired separation (Fig. 4.3).
(2) To facilitate the separation and transportation of paper bags to the bag-opening device, negative pressure suction cups can be used to attach to the surface of the bags [2]. This causes the topmost paper bag to separate from the stack in the storage device, while the suctioned bag can be moved to the bag-opening device by a sliding module for the subsequent opening step (Fig. 4.4).

4.2.2 Micro-Opening Paper Bag

To improve the efficiency of subsequent bag opening, during the transportation of paper bags, the semi-circular notch at the back of the bag and a retractable sheet can be utilized to micro-open the paper bag.

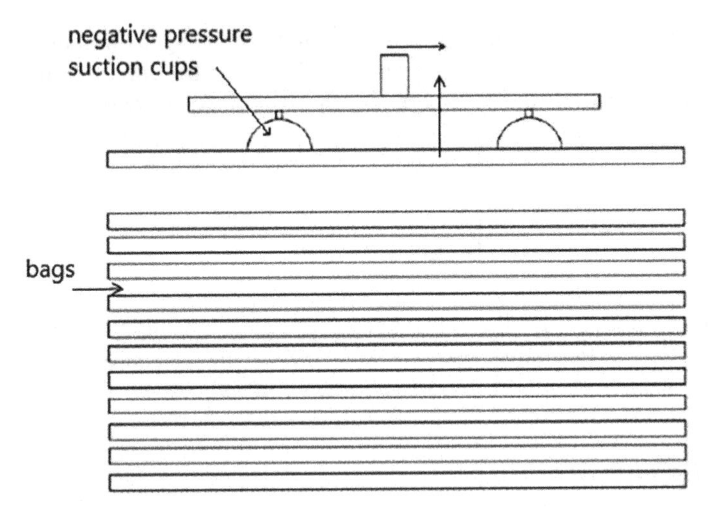

Fig. 4.4 Principle of using suction cups to separate paper bags. Suction cups create negative pressure to stick paper bag on the top and separate it from others

4.2.3 Completely Open the Bag

There are two options to open the paper bag. The first option is to insert a deployable mechanical device into the slightly open bag opening. Once fully inserted, unfold the mechanical device to prop open and shape the paper bag [3] (Fig. 4.5).

The second option involves using two suction cups to hold onto the front and back sides of the bag, creating negative pressure to open the paper bag. Afterward, a shaping device is inserted into the paper bag for shaping, which facilitates the bagging operation [4].

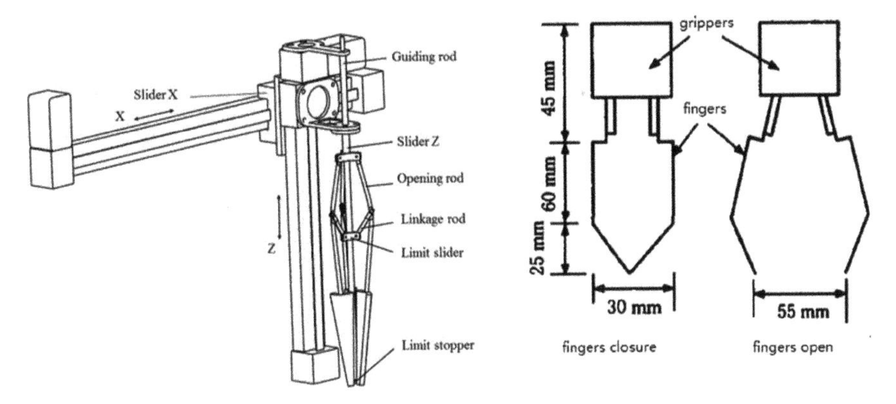

Fig. 4.5 Structure of opening device. Left one use three fingers to open the paper bag, while right one use two to open it. Left one will make opening status steadier

For improved efficiency and effectiveness in bag opening, as well as to simplify the operation process, the paper bag surface friction can be utilized by relying on a roller device to separate and transport the paper bag to the bag opening mechanism. Then, an insertion of the wedge device can fully prop open the paper bag.

4.3 Movement to the Target Position

After identification, move the object to an approximate position without requiring special accuracy. Final adjustments can then be made by the end-effector.

4.3.1 Robotic Arm

(1) The structure of the robot arm should be selected

Based on the requirements for a larger moving path during bagging, a suitable mechanical arm can be selected from those used in industrial production, which can be divided into the following categories (Fig. 4.6).

The right-angle coordinate type [5] consists of three mutually perpendicular moving tracks that change the position of the end-effector by sliding between

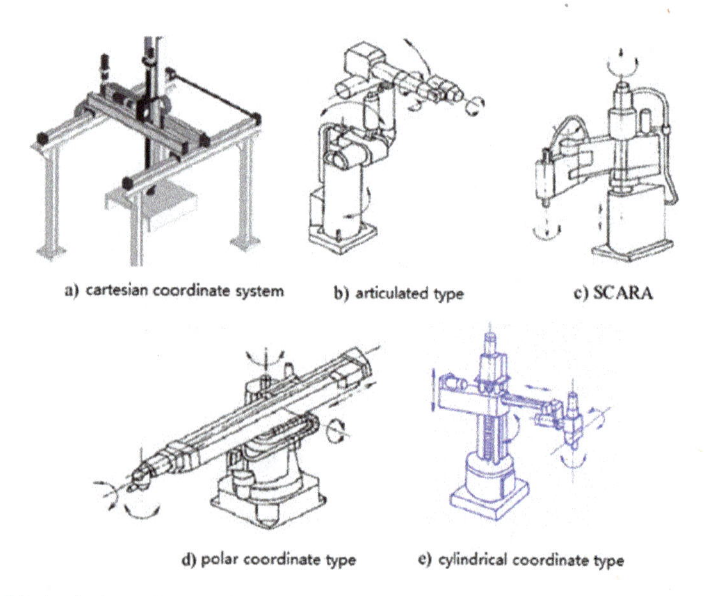

Fig. 4.6 Mechanical arm forms. **a** Photo of Cartesian coordinate type. **b** Photo of articulated type. **c** Photo of SCARA. **d** Photo of polar coordinate type. **e** Photo of cylindrical coordinate type. Additionally, cylindrical coordinate type (in blue) is the one with the most flexibility

each other. However, this type takes up more space, has a smaller range of motion and a fixed end direction.

The polar coordinate type [5] comprises a freely rotatable platform and a rod that can move back and forth and rotate up and down. This type offers high flexibility and occupies a small footprint. However, it may be challenging to operate with significant height differences.

The cylindrical coordinate type [5] comprises a rotating and lifting platform and a rod that can move back and forth. This type offers more flexibility, although the moving joints are susceptible to damage.

SCARA (Selective Compliance Assembly Robot Arm) [5] consists of three parallel rotating joints, offering a lighter and more flexible design. However, this type has a smaller working range compared to other robot arm structures.

The joint type [5] consists of three rotating joints with different axis directions, offering flexible movement and a larger angle adjustment range. Although the movement and control are more complicated, this design is suitable for the project's need for precise positioning, fast operation, and solid obstacle avoidance.

Maintaining good rigidity in materials can be challenging in practice. For this reason, if the robot arm length is too long, gravity may cause bending, making it necessary to keep the arm length as short as possible. The joints should also be as flexible as possible while maintaining a large range of motion to complete all the young fruit sacking.

(2) Motion planning of the robot arm

The positive kinematic solution describes the robot's posture in the reference coordinate system when the parameters of each linkage of the bagging robot are known. Then, using the inverse kinematic solution, the bagging robot can travel along a predetermined route, allowing the robot end-effector to reach the desired posture.

4.3.2 Obstacle Avoidance

Obstacles that the robot arm must avoid include main branches, auxiliary branches, young fruits, leaves, and paper bags on the young fruits. Although leaves are soft and have minimal influence on the robot arm's movement, they should still be avoided. The primary obstacle avoidance targets are the branch trunk and young fruit. Since a multi-joint robotic arm is used, it can avoid obstacles by bending its joint parts. After identifying the target fruit through the camera, a reasonable moving path is planned to complete the bagging operation while preventing mechanical damage.

4.3.3 End Orientation

Due to the high precision required in the bagging operation and the difficulty in achieving absolute accuracy in robot arm positioning, a distributed operation approach can be used. The end-effector is first moved by the robot arm to the vicinity of the target young fruit, followed by the final bagging operation completed using the end moving mechanism.

The bagging process requires strict directionality, requiring the end to have steering and moving functions to ensure accurate bagging by aligning the paper bag opening with the fruit stalk. To achieve a wide range of rotation, two rotating joints can simulate universal joints. Lastly, the end-effector should have a telescopic mechanism to ensure successful insertion of the young fruit into the bag through the bag mouth, completing the bagging process.

4.4 Bag Sealing

Manual bag sealing operations are often cumbersome and difficult to achieve mechanically. To simplify the process, researchers are exploring various changes to the bags themselves. By using different materials and shapes, they aim to achieve better sealing effects and provide customized solutions for different designs.

4.4.1 Stapler Type

Currently, mature product designs on the market mostly use a solution that involves using specially designed envelope-type paper bags. This solution involves placing young fruit into the bag and using a stapler-like sealing device located below the bag's mouth to seal the bag shut. This method is quick, simple, convenient, and more efficient than other methods. However, one drawback is that the paper bags may not be large enough to accommodate larger items.

4.4.2 Squeeze/Engagement Type

This program is based on an improved design of the paper bag used in current manual bagging processes. Generally, paper bags only have a single wire on one side that is used to fix the closed bag mouth. However, this program uses four symmetrically distributed wires instead of just one. When young fruit is placed into the bag, the closure mechanism near the bag mouth is activated. The wedge-shaped mechanism and groove bend the wires to close the bag mouth and secure it shut [6] (Fig. 4.7).

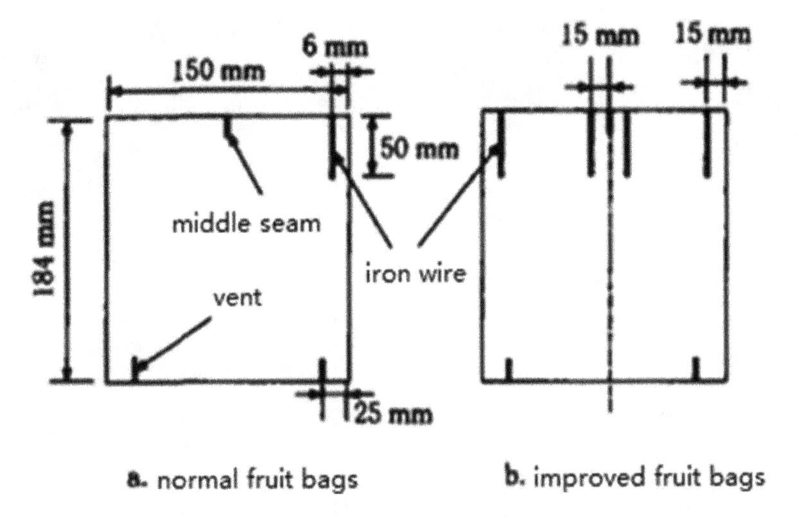

Fig. 4.7 Improvement of fruit bags. **a** Normal fruit bags with only one iron wire. **b** Improved fruit bags with four iron wires. Clearly, improved fruit bags can form a steadier closure

4.4.3 Ring Mouth Shrinkage Sealing

This solution completes the sealing operation by changing the size of the cross-section near the bag mouth. This can be achieved in two ways. The first way uses two fork-shaped mechanisms that move relative to each other, causing the enclosed cross-section to become smaller and completing the sealing operation. The second way uses a topological linkage mechanism which is simpler in principle but may result in poorer sealing compared to other methods. It is also more prone to loosening and cannot effectively prevent rain from entering the bag [7] (Fig. 4.8).

4.4.4 Hot Melt Adhesive Sealing

To achieve a more effective seal, it is possible to apply or add hot melt material to the mouth of the bag. Afterward, both sides of the bag are tightened and heated using a heating piece to melt the material and complete the sealing process. While this approach yields better results, the heating process may damage the fruit stalks and increase production costs. Additionally, removing the seal later could be more difficult.

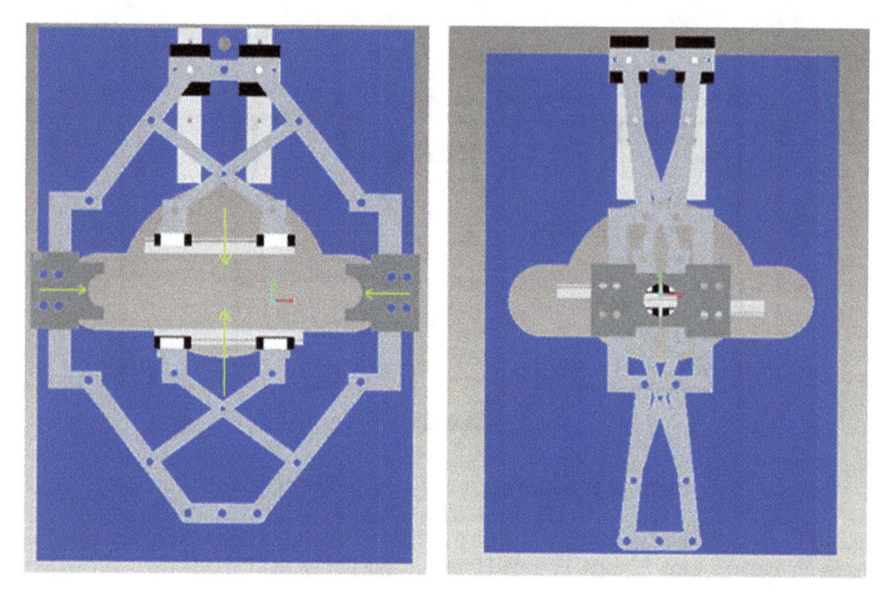

Fig. 4.8 Structure of sealing device. The inside mechanic structure creates push to seal the paper bag tightly

4.4.5 Non-Paper Bag Sealing

To seal plastic film bags, young fruit is inserted into the bag and the rod drive is used to close the upper and lower bag mouth. The bag mouth is then heated until melted and adhered together to achieve a proper seal. For other bags with slot designs or boxes, they can be sealed by nesting their slots together [8].

Considering the need to adapt to current production demands for more cost-effective bagging, achieve better seals, and facilitate subsequent bag removal, the preferred option is to use circle mouth shrinkage for bagging double paper bags [9] (Fig. 4.9).

4.5 A New Type of Apple Young Fruit Bagging Robot

The new type of apple young fruit bagging robot mainly comprises four parts: a control system, detection device, drive device, and actuator.

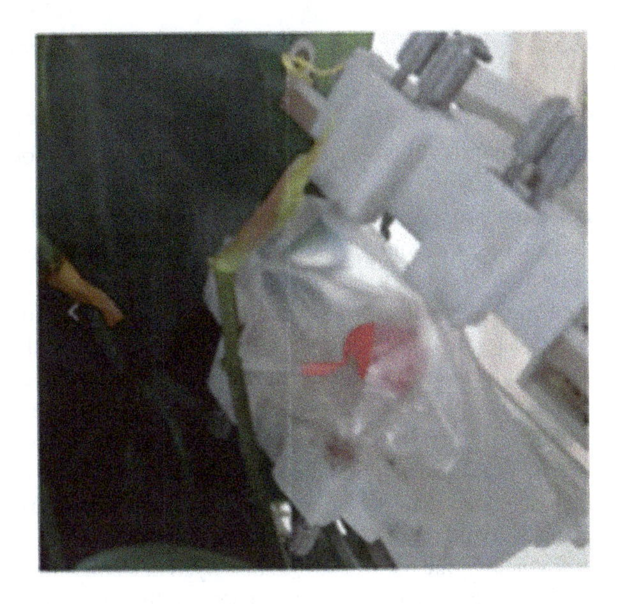

Fig. 4.9 Common non-paper fruit bag. Plastic fruit bag, which can be shaped into many different forms and be much easier to cover the young apple, will make whole process highly efficient

4.5.1 Control System of Apple Bagging Robot

The control system's core consists of a programmable logic controller (PLC) and a microcontroller. The PLC can perform logic operations, sequential deployment control, timing signal triggering, counting, and arithmetic operations, among others. Its high reliability makes it suitable for process control in large-scale mechanized production but comes at a higher cost. On the other hand, the microcontroller finds use in smaller control systems and being a small but complete computer system, still offers powerful functions at a lower cost. As a result, it is more widely used in various fields of daily life.

Since the control variables in the apple fruit bagging robot's control system mainly involve the start and end of the action, a microcontroller is more appropriate for use. In designing the control system, it is essential to predefine corresponding parameters such as the time points, degree of motor forward and reverse rotation, trajectory of the robot arm movement, and more. In terms of transporting paper bags to the bag-holding area, it is primarily accomplished through controlling the start and stop times of the bag delivery roller's rotation. When using the vacuum negative pressure method to open bags during bag propping, the control system requires an electromagnetic changeover valve to control the cylinder's start and stop. If the pure mechanical method is selected instead, the trajectory and process of the mechanical bag opening device must be accurately compiled to determine the precise moment when the device opens and props up the bag. During the sealing process, it becomes necessary to control the wedge-shaped sealing mechanical device to exert an equal driving force in the opposite direction, bending the built-in wire of the paper bag mouth into a target shape to seal the bag's mouth [10].

4.5.2 Apple Bagging Robot Detection Device

The apple young fruit bagging robot's detection device primarily comprises a vision system that incorporates a camera, a microcontroller for processing image information, and a software environment for compiling recognition algorithms. During the detection process, it is necessary to collect images that have sufficient resolution and analyze the young fruit features in these images through machine vision finding algorithms to find the target young apple fruit. Once found, measurements are taken by a binocular stereo vision system to determine the young fruit's size and its relative position to the robot arm's stalk. This series of data is then transmitted to the control module. In the event of an obstacle, it is necessary to execute the planned obstacle avoidance action. This involves calculating the distance from the robot to the obstacle and determining the best route based on image information that allows the robot to bypass the obstacle and reach the bagging operation position. The control system then sends a signal to the drive motor to steer and navigate the robot and achieve obstacle avoidance.

Typically, the obstacle avoidance system's software design uses an ARM + DSP dual-core internal communication design that facilitates data interaction by sharing data between two CPUs. To ensure the shortest path and the least consumption in the obstacle avoidance process and develop the correct obstacle avoidance strategy, one party sends an interrupt signal to prompt the other party to receive data. The receiving party then enters the interrupt service program to complete data reception. During the vision processing stage, the most critical indicators are the resolution of the acquired image, the calculation speed of collected data, and the accuracy of the recognition process. These must be optimized and improved multiple times [10].

4.5.3 Driving Device of Apple Sleeve Robot

The drive device serves as the core device for system output actions and is mainly divided into three types: pneumatic, motor, and hydraulic drives. Given the apple young fruit bagging robot's small size, a small and convenient motor drive system is the most suitable choice for the robot's drive device and energy supply. The small motor is primarily installed in the bag delivery roller, mechanical bag support device, and wedge sealing mechanism, allowing the control system to carry out forward and reverse operations that promote the realization of each output action.

4.5.4 Actuator of Fruit Bagging Robot

For bagging operations, a multi-joint mechanical arm is selected. Firstly, the bag storage device is combined with the bag-opening device. After separating the single paper bag, the wedge-shaped claw props up the bag entirely. The mechanical arm then grabs the propped-up paper bag and fixes it to the end actuator. Meanwhile, the vision system identifies the young fruit on the tree and feeds its position to the control system through binocular vision. The path is planned, and obstacles are bypassed. Afterward, the robot arm moves along the planned path to the vicinity of the young fruit. The end-effector makes the final positioning and positions the young fruit to enter the bag. Finally, the bag is sealed by engagement.

4.6 Similar Listed Products

Three companies in the country that provide apple bagging services have been identified: Yi Zhongli, Good Fruit, and Fruit Easy Set. After a period of website research, an analysis of available information on these companies will be conducted by the author to arrive at an objective assessment.

4.6.1 Yi Zhongli Company

You can find a lot of video information on this company's official website, which informs that the company's listed products mainly focus on the apple bagging process of opening the bags. Paper bags are stored in the bag storage bin in the back and then come through the single-chip drive bag delivery mechanism, bag support mechanism, bottoming mechanism, to complete the opening of paper bags. Four symmetrical pairs of suction cups are evenly placed on the upper part of the bag. During the operation, four pairs of suction cups grip the paper bag. Then the symmetrical suction cups move apart to open bag with negative pressure. However, because of the two-layers structure inside the paper bags, it may cause the situation that there is only one layer on the one side but three layers on the other, which is a main problem of the product (Fig. 4.10).

4.6.2 Good Fruit Company

The company's official website could not be found on the internet, but information related to the company's products was seen in a video demonstration. It was observed that the company, which is also known as Yi Zhongli, has the main focus on the direction of the bag opening products. There are three different products developed by the company to open and store the bags in a stacked configuration without any bag transfer system. The only difference between these three products is in the way the bag is opened. The first product uses a strong wind blowing method to open the bag by blowing air out of a small hole and achieving the purpose of propping the bag.

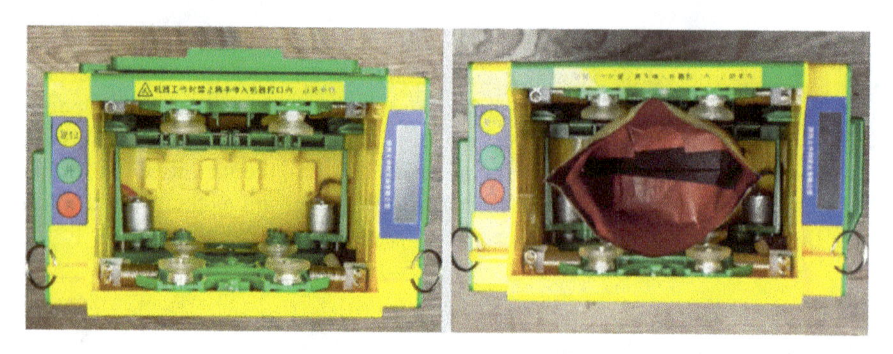

Fig. 4.10 Inside structure of Yi Zhongli Paper Bag Opening Machine. It is mainly based on eight suction cup to form negative pressure to open the paper bag

However, this method may face issues with bags that have multiple layers on both sides, leading to inaccurate blowing and bag displacement. The second product uses pressure to deform the fruit bag half-moon gap through the application of rubber bands, enabling bag opening. Although this method accurately separates the inner and outer layers of the bag, it cannot maintain consistent pressure with reducing bag size, resulting in reduced bag openings. The third product uses a retractable propping mechanism in addition to the first blowing method. First, the bag is blown open and then expanded using the retractable mechanism to fully support the bag. Nonetheless, this product faces the same shortcomings as the first blowing method (Fig. 4.11).

4.6.3 Fruit Easy Set Company

The least relevant information was found about this company, with only a concept video related to their product being available. From the video, it can be inferred that the machinery is an integrated machine that props, sends, and sets fruit bags. The design's unique use of special fruit bags simplifies bagging problems significantly. However, due to the limited availability of these specialized bags, the popularity of the product may be hindered.

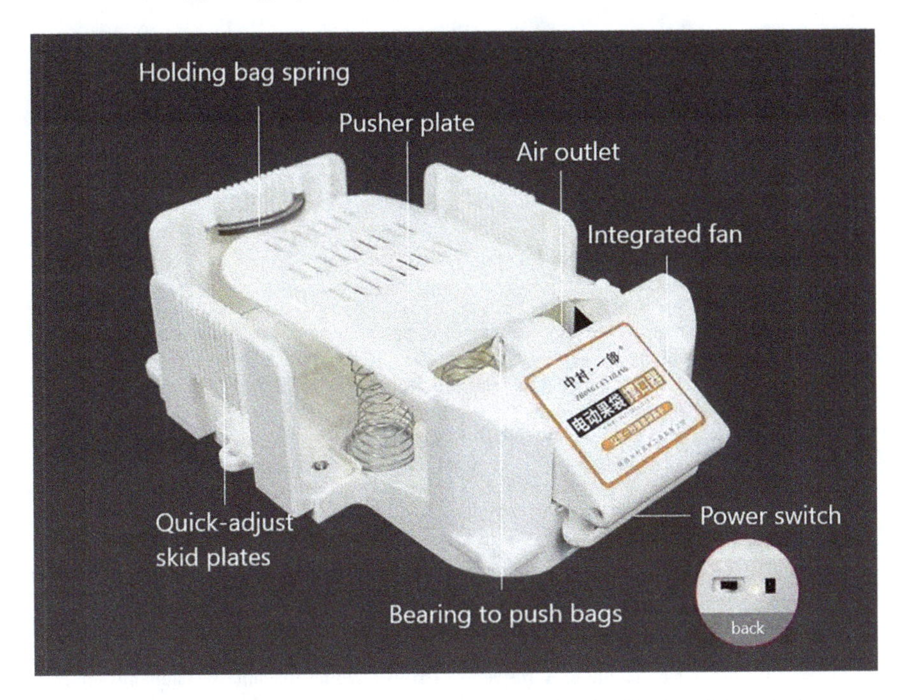

Fig. 4.11 Good Fruit Bag Opener. The opener opens paper bag by blowing air into the bag

4.7 Conclusion

In this study, the main process of existing apple fruit bagging technology was discussed, including several steps, such as visual locating, mechanical bag opening, and bag sealing. A comparative analysis was conducted to evaluate the advantages and disadvantages of different solutions for each step, with a focus on identifying future research goals and directions for apple fruit bagging.

Currently, there are numerous achievements in vision technology, with advanced image processing algorithms, target recognition, 3D reconstruction and depth optimization technologies having higher success rates for function realization. High resolution machines and smooth communication between them have further enhanced these results. However, in the subsequent bag opening and sacking process, most existing cases are still theoretical, with less actual study of the physical finished product due to the high precision requirements for mechanical operation. Additionally, the majority of existing sealing mechanisms do not use commercially available paper bags, resulting in a lack of universality. This has led to an increase in the cost of single bagging, posing a challenging problem to solve.

In summary, there is significant potential for the development of apple young fruit bagging technology in the future. Further exploration, development, and research are necessary to achieve automation of this process, which can reduce the burden on farmers and increase their fruit income. However, due to the author's limited expertise, errors and omissions may exist in the collation and combination of existing technical information. Therefore, readers are welcome to provide corrections or additional information.

References

1. Zhe L (2014) Research of the key techniques of apple automatic bagging based on machine vision. Shenyang Ligong University, Shenyang, MS Thesis, pp 13–29
2. Guang L (2020) Design of a bagging machine for young apple fruit. Chengdu University, Chengdu, MS Thesis, pp 15–20
3. Xia H, Zhen W, Chen D, Zeng W (2020) Rigid-flexible coupling contact action simulation study of the open mechanism on the ordinary multilayer fruit paper bag for fruit bagging. Comput Electron Agric 173(105414):3–4
4. Liming X, Xiaotang G, Tiezhong Z (2007) Automatic unfolding-opening mechanism of fruit bag. Transactions of the CSAE 23(4):139–140
5. Shenghui W (2016) The development design and research of a fruit bagging robot. Yanshan University, Qinhuangdao, MS Thesis, pp 16–23
6. XiaoTang G (2005) Study on semi-automatic robot for apple bagging. China Agricultural University, Beijing, MS Thesis, pp 13–14
7. Oupeng F (2011) Key technology research of fruit bagging machinery. Zhejiang Sci-Tech University, Hangzhou, MS Thesis, pp 12–13
8. Wenyu Y, Yuhong Z, Pan W, Xiangdong L (2019) Development of handheld continuous bagging machine for fruitlet. Agric Eng 9(08):6–9
9. Xiaoping G, Wanjun Z, Jingxuan Z, Jingyi Z, Jingyan Z (2019) Study on the structure design and feasibility analysis of apple inhaled box bags based on hailproof. IOP Conf. Ser Earth Environ Sci 440,022048:4–5
10. Yameng B, Delin Z, Yuan C (2021) Fruit bagging robot. Hebei Agric Mach. https://doi.org/10.15989/j.cnki.hbnjzzs.03.020:38-39

Chapter 5
Sensing and Automation Technologies Applied in Pollination

Meiwei Li, Afshin Azizi, Zhao Zhang, Dongdong Yuan, and Xufeng Wang

Abstract With the expansion of orchards and market demand, the need for artificial pollination is increasing. Manual pollination is a laboring and dangerous task. As the orchards have faced labour shortages and increasing labor costs, researchers are turning their attention to automated pollination robots. Aerial pollination uses Unmanned Aerial Vehicles (UAVs) to pollinate crops in two directions: wind field and precision. Wind field pollination is achieved by blowing pollen into the air through the wind field generated by the UAV during flight, thus achieving wind-borne pollination in natural pollination. Aerial precision pollination is similar to ground-based precision pollination in that a navigation system such as GPS with a navigation algorithm are used to plan the route, or incorporate a camera for route planning assistance. The camera takes pictures of the flowers, identification and location them, and determines the status of the pollinated flowers through CNN-based algorithms. The chapter describes in detail the automatic navigation techniques and image recognition techniques used in the precision pollination process, illustrating the entire pollination process by pollination robots. It concludes with a discussion of existing issues and an outlook on future directions.

M. Li · A. Azizi (✉) · Z. Zhang
Key Laboratory of Smart Agriculture System Integration, Ministry of Education, Beijing 100083, China
e-mail: azizi@cau.edu.cn

Key Laboratory of Agricultural Information Acquisition Technology, Ministry of Agriculture and Rural Affairs, China Agricultural University, Beijing, China

College of Information and Electrical Engineering, China Agricultural University, Beijing 100083, China

Z. Zhang
e-mail: zhaozhangcau@cau.edu.cn

D. Yuan
Sweet Fruit, Co., Ltd, Suqian 223839, Jiangsu, China

X. Wang
College of Mechanical and Electrical Engineering, Tarim University, Alar 843300, China
e-mail: wxf@taru.edu.cn

Z. Zhang and X. Wang (eds.), *Towards Unmanned Apple Orchard Production Cycle*, Smart Agriculture 6, https://doi.org/10.1007/978-981-99-6124-5_5

Keywords Precision pollination · Wind field pollination · Flower identification and localization

5.1 Introduction

Pollinator-dependent plants provide medicines, biofuels, pigments and fibers for human life [1]. A study [1] has indicated that a correlation exists between pollinator crops and the dietary requirements of individuals in certain geographical zones. For instance, regions with elevated levels of vitamin A insufficiency display a tripled dependency on pollinator-reliant crops as a source of plant-derived vitamin A in comparison to other regions. If there is insufficient pollination, it may cause a decrease in the global crop yield by 5–8% [1]. However, expanding farmland comes at the expense of forests and grasslands, and also reduces habitat for pollinators [2]. Growers have resorted to using significant amounts of pesticides due to the extensive utilization and management of land. This heavy reliance on pesticides has been identified as a major contributor to the abrupt reduction in pollinator populations [2]. In areas where natural vegetation is well protected, pollinators are abundant and do not require artificial pollination. The sharp decline in the number of pollinating insects, as a result of the degradation of the global ecosystem, and the increasing monoculture of economic plants, which raises the incidence of pests and diseases, coupled with an exponential increase in market demand, imply that natural pollination alone is insufficient to meet commercial production requirements. Some European countries have recorded that over 50 species of pollinating bees are currently under threat of extinction [1]; and the large number of bees breeding can easily lead to the spread of crop diseases and parasites. The foraging behavior patterns of honey bees vary depending on the crop species. Their promotion of cross-pollination notwithstanding, they can result in repeated pollination which could render the quantity and quality of pollen delivery unstable, leading to an inconsistent fruit set [3]. Natural pollination is strongly influenced by environmental changes, such as low temperatures, and reduced pollinating insect activity, which can result in subsequent yield reduction. Or high temperatures affecting pollen vigour leads to flowers without fruit development [4]. Hence, the commercial orchards cannot afford these instabilities.

In order to increase crop yields and prevent excessive fruit deformities in naturally pollinated crops [5], artificial pollination is now applied alongside natural pollination. The practice of artificial pollination, which emerged in China in the 1990s as a means to increase crop yields, is labor-intensive and economically unsustainable due to the high cost of labor involved. During manual pollination, workers may inadvertently damage the stigma, which can result in uneven pollination force and ultimately lead to reduced yields [6]. The flowering period of the crop usually varies from 1 to 15 days, and failure to pollinate flowers in time can directly affect fruit set. In addition, workers will come into direct contact with pesticides such as insecticides and herbicides during the pollination process, and two-handed spotting can lead to safety hazards when pollinating from high places. Growers have therefore invented various tools to increase pollination efficiency. Examples of such tools include manual liquid pollination sprays and electric vibrating pollinators for kiwifruit pollination [6]. It is worth noting that these tools are designed for one-handed operation, reducing the workload on workers and increasing efficiency.

Large and well-contacted pollinators, such as bees, can provide better pollination results and floral resources [1]. Although hand pollination can fulfill most of the yield requirements, labour costs have increased due to the loss of young workforce from orchards, caused by rapid technological development and accelerated urbanization. Artificial pollination is inefficient compared to machines. A research has shown that using unmanned helicopters for pollinating hybrid rice yields comparable results to hand pollination [7]. To pollinate apple trees, each worker can pollinate only 5–10 trees per day. In contrast, UAVs equipped with several sprinkler heads can pollinate an entire orchard at a maximum operational flight speed of 7 m/s [8]. In addition to aerial pollination using UAVs, there are also pollination robots that perform small-scale precision pollination on the ground using robotic arms and nozzles. Another popular application of pollinator robots is for kiwi pollination, where ground-based robots are used. Kiwifruit is the largest horticultural export in New Zealand. New Zealand has been conducting ongoing research on kiwifruit pollination robots, which has been documented in several studies [9–14]. One of the main techniques is to identify the flower and select one of the nozzles in a row to spray a pollen solution on the flower to complete pollination. The use of plant hormones for pollination has been explored in research [15], which demonstrated the real-time identification and pollination of a tomato plant with a sprayer at the end of a single robotic arm. However, most growers in Mao County believe that if they insist on growing crops that require pollination, artificial pollination is the only solution [2]. As a result, farmers in Mao County are abandoning hand-pollinated crops such as apples and replacing them with non-pollinated crops such as vegetables, plums and loquats [2]. Therefore, highly practical and low-cost automatic pollination robots are urgently needed.

This chapter aims to illustrate the technique of automatic pollination. The second section introduces air pollination and its technical application. The third part introduces ground pollination, that is, pollination technology using mechanical arms and other facilities. Finally, the application limitations and future development of pollinator robots are discussed.

5.2 Wind Field Pollination

In greenhouses, crops need to be pollinated artificially due to the enclosed nature of the greenhouse. Figure 5.1 by [16] demonstrates a straightforward method for pollinating crops without the need for human labor. This involves attaching deflectors to a UAV mounted on a tripod, which obstructs the downdraft and generates a wind field parallel to the ground. This process effectively completes the pollination work. At a height of 1.5 m, the UAV generated a wind field that reduced the wind speed from

Fig. 5.1 The design of drone from [16]

10 m/s to 7 m/s. The deflectors attached to the UAV also helped to spread some of the wind horizontally, resulting in an increase in the horizontal wind speed. The research [16] concluded that this solution is feasible for the pollination of self-pollinated crops such as tomatoes in greenhouses.

To address the labor-intensive and inefficient nature of artificial backpack pollination, plant protection drones have been used in outdoor pollination trials with small unmanned helicopters [8, 17, 18]. In a research [8], a DJI T20 plant protection multi-wing drone was used for aerial pollination of pear trees. Although the power of plant protection UAV can usually only support 10–40 min of flight time, compared to the worker who pollinate 1000 flowers a day, the UAVs greatly save pollination time with the flight speed of about 3 m/s [18], and is not restricted by manual working time.

After loading the pre-configured pollen nutrient solution into the pesticide tank, the T20 was flown at 3 m above the ground to carry out aerial pollination operations with a maximum aerial speed of 7 m/s [8]. The results have analyzed using a five-point sampling method and water-sensitive paper to obtain droplet deposition distributions. The effect of these five parameters on droplet deposition is investigated under slightly different natural conditions, based on different flight speeds, flight trajectories, flight altitudes, liquid nozzle sizes and spray volumes. The flight trajectory was divided into two parts: flying over the top of trees and flying between the trees. The results showed that droplet deposition was more effective when flying above the canopy than when flying between the trees. As the flight height increased, the droplets became

Fig. 5.2 Image of UAV flying between trees from [18]

more sensitive to changes in wind speed and direction, and the density of droplets deposited at the sampling sites decreased (Fig. 5.2).

Aerial pollination is flexible and has advantages for orchards with high canopies and complex terrain. Compared to artificial and liquid pollination methods, UAV pollination requires less pollen usage, resulting in cost savings of US$12.69/667 m^2 and US$3.32/667 m^2, respectively. Additionally, UAV pollination is significantly faster than manual shaking and liquid pollination, with timeliness that is 40 and 20 times greater, respectively [18]. However, plant protection multi-wing UAVs have a very limited load capacity due to their energy supply and consumption limitations. Typically, they can only carry pesticide volumes ranging from 20 to 100 L. Thus, these UAVs adopt ultra-low volume spraying technology. The spray volumes per unit area are lower than those of backpack sprayers, and aerial operations leading to a concentrated distribution of droplet deposition in the upper canopy layer [8], where the density of droplet deposition is one of the factors determining yield [18]. Both research studies [8, 8] concluded that the fruit set rates from UAV pollination were lower than those from manual spot and backpack pollination [8].

Agricultural plant protection drones, also known as single-rotor drones, are designed for use in crop protection. In addition to multi-wing drones, which can carry pollen nutrients and pesticides, single-rotor drones have a special parameter—the downwind field. Agricultural drones usually work 1 m to 10 m above the crop [18, 19]. The downward rotating winds generated by the drones' wind field will directly affect the crops. The wind generated by the UAVs act on the pollen in two forms, one on the stem and the other on the pollen. The shaking motion of the stem dislodges the pollen, while the wind blows it away [20]. The research studies [7, 17, 19, 20] utilized monoplane drones to pollinate hybrid rice, taking advantage of their high wind power and fast flight speed. Hybrid rice has a short flowering period, lasting only 1–2 h per day, and the overall flowering period spans 7–10 days. Additionally, its pollen can only survive for 4–5 min, making it unsuitable for use in stored or pre-configured pollen solutions. Compared to manual rope pollen catching and single or double pole pollen catching, monoplane drones are capable of spreading the pollen of parent plants over greater distances, thus allowing for an expansion in the row ratio and compartment width between the seed-making parents [17]. The heterosis fertility rate of drone-assisted pollination is above 42% and the yield is around 200 kg/667 m^2, which is similar to that of hand-pollinated hybrid rice in the same year [17]. Directly observing pollen distribution under wind fields can be challenging. To address this, a study [19] proposes that the pattern of rotor wind fields could serve as an indicator for pollen distribution, particularly under ideal conditions.

To acquire data on the wind field, the researchers utilized a variety of tools, including GPS to track the UAV's flight parameters, an array of ground-based wind speed sensors, and a wind speed data receiver [19]. At each measurement node, three wind speed sensors were placed to measure different axes [19]: the X-axis, representing the direction of flight; the Y-axis, which runs parallel to the ground but perpendicular to the X-axis; and the Z-axis, which is perpendicular to the ground. The wind directions X and Y, which are parallel to the ground, play a significant role in the suspended transport of pollen, while the Z direction mainly affects crop stability. Lower wind speeds in the Z direction can be beneficial for crops.

The monoplane UAV flew along the male parent row at a height of 1 m, maintaining a speed of 3 m/s [19]. It continued until reaching the end or front end of the row, where it then hovered to wait for the next test to begin. According to the results [19], wind speeds that are truly favorable for pollination, i.e., wind speeds greater than 1 m/s in the X and Y directions, are not widespread, typically ranging between 3-5 m. While the Z direction has the maximum wind speed, its effective wind field width is the smallest, at about 2 m. Research [20] conducted a test on the flight speed parameter, and compared the resulting wind field width with the results from previous research [19]. The system selects three different speeds—3.5 m/s, 4 m/s, and 4.5 m/s—and relies on the BeiDou Navigation Satellite System (BNSS) for positioning. To obtain the distribution of pollen quantities, a slide coated with vaseline was placed at sampling points spaced 1 m apart, as shown in Fig. 5.3. The speed of 4.53 m/s was the most favorable flight speed for pollen dispersal. The pollen counts at the left node were significantly greater than pollen counts at the right node, indicating an asymmetric effect of the UAV wind field on pollen distribution.

Another pollination is between wind field pollination and precision pollination [21]. It making pollination solution becomes bubbles solution and pollinated by a UAV. To promote pollen germination and floral tube growth, commercially available surfactant [A-20AB] is mixed with pollen at a rate of 4 mg/L along with H_3BO_3, $MgSO_4$-$7H_2O$, $CaCl_2$, and KCl. The solution is placed in a bubble maker, attached to the underside of the main body of the UAV, pollinated at an angle of 70°–80° and

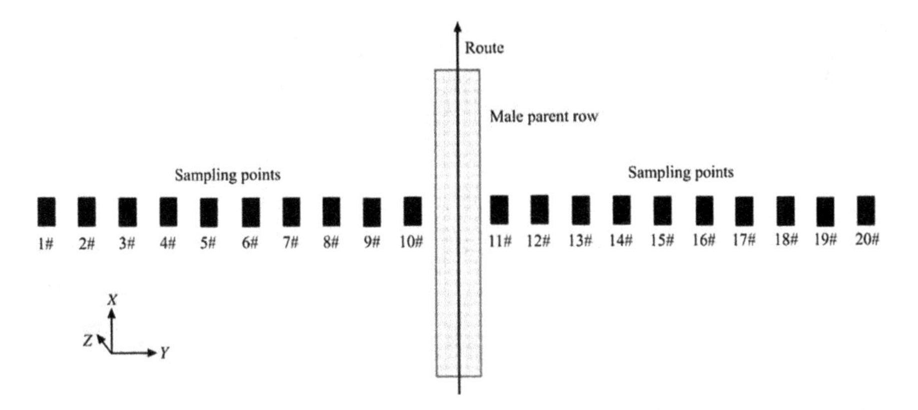

Fig. 5.3 UAV flight path planning

at a height of about 2 m from the ground towards the flowers. The high viscosity and softness of the bubble allows the pollinating bubble to stick firmly to the pistil and avoid mechanical damage from contact pollination. Pollen is exclusively distributed on the liquid film surrounding the bubble, preventing non-specific flower pollination from scattered pollen. This helps maintain fruit set rates and reduces the need for excessive flower and fruit thinning work. During their experiment, Yang and Miyako [21] added 2% HPMC (hydroxypropyl methylcellulose) to the bubbles in order to increase their mechanical stability and ensure smooth pollination under the wind field of the UAVs. The results showed that the bubbles would not rapidly dissipate when stationary on the contact surface for at least 10 min. Although the experiment achieved a successful pollination rate of over 90%, this was only for a single instance of pollination. However, on a larger scale such as a farm or orchard, many bubbles may miss the flowers resulting in wasted pollen. Moreover, it is important to note that bubbles used for pollination can be mildly toxic, and if there is a large amount of wasted bubbles, it could potentially lead to toxic deposition and contamination of the farmland. As the number of bubbles received increases, the amount of pollen deposited in the pistil also increases. However, after accumulating 10 bubbles, many flowers fail to pollinate due to toxic deposition, potentially leading to reduced yields. The activity of pollen in solution remains to be determined, and the labile activity of pollen makes bubbles unsuitable for most crops, leaving a long way to go in terms of large-scale cultivation (Fig. 5.4).

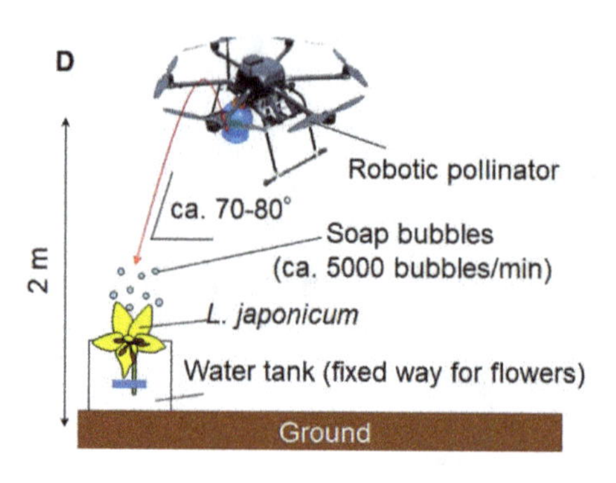

Fig. 5.4 Schematic illustration of pollination by drone from [21]

5.3 Precision Pollination

In addition to the use of wind farms for large-scale pollination, single-plant pollination using precision pollination has been widely investigated as another direction for automatic pollination. The focus of the technology is on enabling the robot to automatically plan its path through orchards or greenhouses before pollination, orient itself by identifying clusters of flowers, and accurately identify and locate pistils for effective pollination. The pollination robot repeats the process of identifying and locating the pollination clusters, recording the flower clusters and predicting the time of flowering in order to improve the fruit set rate. Robotic pollination offers distinct advantages over artificial pollination by enabling the selection of specific flowers at various stages of growth through cluster identification and recording, thereby reducing the need for labor-intensive fruit thinning and improving overall fruit quality [22].

In addition, small robots developed with bionic technology can be used to perform precise pollination tasks [23], but the price of manufacturing and application of such small bionic robots is expensive, and the cluster communication function of robobees has yet to be realized for pollination purposes. The bionics knowledge is also used to imitate the natural pollination of bees with soft and fine belly hairs [24]. Ionic liquid gels with attached animal hair are positioned beneath the main body of the drone. The drone is then controlled to mimic bee behavior and pollinate the female flower by touching the part of the ionic liquid gels that contain pollen particles. The research [24] focuses on the physical and chemical properties of the ionic liquid gels material itself, as well as its stability and ability to effectively adsorb pollen particles under different environments, rather than automatic pollination.

5.3.1 *Route Planning*

- Small drones

For aerial pollination by small UAVs and unmanned helicopters, route planning for aerial pollination can be set in advance because the aerial view is significantly wide and the UAVs can navigate by GPS or BNSS. The hardware can perform remote control and record flight path functions, so that the recorded path data can be returned to the control side by wireless transmission. However, the automatic route planning capabilities of most small UAVs are limited to simple tasks such as planning straight lines, U-shaped routes, and making minor adjustments to straight lines [25]. When it comes to more complicated route planning, manual manipulation is often required. The route planning in [25] uses the DJI Phantom 4 for image recognition of the male OSR (regarding to the hybrid oilseed rape) the least-squares method (LSM), and Hough transform to fit a distributed route to the male OSR for planning the flight path. However, to obtain high quality captured images, it requires manual adjustment of the UAV position and lens angle by the operator to trigger camera shots. In research [26], a Maix Bit processing board hardware was mounted above the central torso position of the DJI Tello drone. The first step of navigation is taken by a color camera mounted above the board, which utilizes a deep learning based image recognition with a PID control loop to execute the navigation algorithm until the UAV is less than 0.8 m within the flower. The second step of navigation is within 0.8 m of the flower, based on the IBVS (Image-based Visual Servoing) end-to-end network, supplemented by a ToF distance sensor to detect and approach the flower for final pollination. Image recognition-based UAVs have a common drawback: once the recognition target exceeds the camera's range or experiences a slightly faster displacement due to natural wind or high-intensity light interference during filming, it becomes difficult to navigate the UAVs to their intended pollination destination. So the target position or camera angle needs to be readjusted and the image needs to be re-taken and recognized. That is a challenge for navigation in natural environments.

- Robot arm

For multi-degree-of-freedom robotic pollination, the operating space of the robotic arm can compensate for computational errors in the path, supplemented by radar, sensors and cameras to avoid obstacles between paths. The robotic arm also has more space on the operating platform than the drones and can be configured with a wider choice of hardware. This makes it a more versatile option for tasks that require complex or specialized equipment. The path can be controlled remotely by workers [15]. Alternatively, red navigation lines can be laid on the path, and video cameras can be used to capture real-time information about the navigation lines and detect obstacles. This allows the pollinator robot to follow a preset red route [15]. 3D Lidar in research [27] is used to scan the indoor space and match it with the 3D indoor map prepared in advance. If it exceeds the set threshold, the matching succeeds, and the pollinator is added to the factor graph with the current attitude. The 3D SLAM

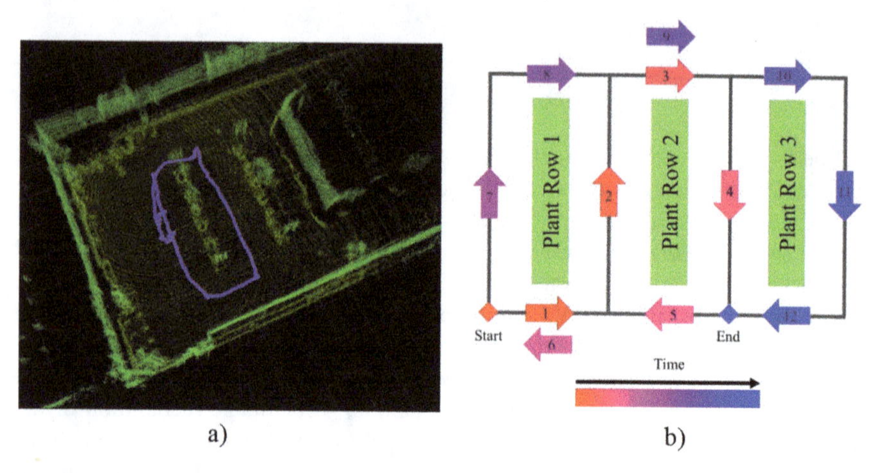

Fig. 5.5 **a** 3D spatial scan generated by SLAM, with the robot's path indicated by the blue line; **b** displays the Voronoi path planning diagram, where the time taken to move from the red to the blue point represents the path between those two points

map employs the Generalized ICP algorithm and GPS system to estimate the offset of the robot's initial attitude relative to the origin in global coordinates. Additionally, the map utilizes sensor data fusion to generate a more precise and efficient attitude estimation for the robot's actualized route. The pollination robot will survey the greenhouse and create a Voronoi diagram to establish a detection path. During this time, a fish eye camera observes its surroundings, and follows the planned path automatically. To achieve the final pollination, a depth camera will be installed in the robot and visual servoing techniques will be employed. A large pollinator robot [9] uses the ROS platform for route planning. Research on cluster communication for swarms of bee bionic robots is still ongoing [23], while ultrasonic pollinators are currently fixed in place and vibrate the air to pollinate crops row by row [28] (Fig. 5.5).

5.3.2 Pollination Preparation—Flower Identification

Both UAVs in studies [26] and [25] utilized a color camera to capture images and a CNN to construct an image recognition model. In the study [25], the entire image recognition process takes an average of 1.18 s in cloudy weather conditions that are favorable for distinguishing male OSR plants. Additionally, in study [26], a successful fully autonomous navigation pollination process takes approximately 31.5 s.

Studies [5, 15, 27, 28] have utilized binocular stereo depth cameras for object recognition and localization in precision pollination robots with robotic arms. These cameras convert the target to the actual coordinates to the camera coordinate system based on the principle of human parallax to determine the specific location information of the object [29]. After capturing the actual object image by the binocular camera, the image needs to be processed and identified. Prior to processing live images, researchers should first train the image dataset to allow the data processing center to quickly identify the flower after the camera captures the image.

Image datasets consist of large amounts of data that have been generated through data augmentation of pre-existing photographs [30]. These datasets were subjected to deep learning model training based on different algorithms selected for different studies. In research [15], a traditional threshold segmentation method was used to numerically analyze the H (Hue), S (Saturation), and I (Intensity) values images to detect the presence of a flower cluster. Research [30] argues that this traditional threshold segmentation method is flawed and has a low accuracy for nighttime or under cloudy conditions. In [31], researchers argued that YOLOv4 outperforms YOLOv3 in detecting small targets, such as flower buds. This approach was used to accurately predict the blooming peak by counting the number of flower buds. Research cited in [31] shows that YOLOv4 has a much faster detection speed compared to Faster R-CNN (1833 ms per image [9, 10]) and Mask R-CNN (372 ms per image). Specifically, YOLOv4 can accurately detect objects in only 38.64 ms per image. In [32], researchers tackled the challenge of estimating the orientation of flowers for accurate pose identification by a pollination robot. Although back-projection localization of camera coordinates can identify the flower, it cannot determine its orientation directly. To address this issue, the researchers classified the orientation of each flower, enabling the robot to accurately estimate its pose and validate the identification. The estimate pose classifier was validated with an accuracy of approximately 70%. Researchers in [27] designed an endoscope camera intended to aid the end-effector motion of a robotic arm. However, this camera was found to have limitations in its ability to observe objects at close proximity. Researchers in both [31] and [30] showed the importance of early detection of pollinated flowers for future pollination efforts. In [30], center flower identification accuracy was compared between YOLOv4 (which detects buds) and the Mask R-CNN algorithm (which detects the earliest center). Results showed that at the 20% flower stage, the accuracy was much higher (98.7%) than at the 80% flower stage (65.6%).

The challenge of image recognition is exacerbated by natural illumination, which can obscure texture and color boundaries under bright light. For small drones, high winds add another level of difficulty.

5.3.3 Pollination Process

Due to the challenge of controlling the speed and precision of small UAVs, the bionic bee pollination behavior described in [24] is likely to cause mechanical damage to pollinated flowers. In research [26], an 8 cm pollination rod was attached to the UAV, and successful pollination occurred automatically when the distance between the UAV and the sunflower was less than 8 cm. However, merely touching sunflower flowers is unlikely to result in successful pollination. Although the bubble method described in [21] does not damage flowers as much as methods in [24, 26], it lacks the autonomy needed for precise navigation during pollination. After identifying the location, successful and precise pollination remains a challenge for UAVs.

Compared to small UAVs used for precision pollination, there are numerous options available for pollination tools that can be installed on a robotic hand, which is much larger in size. In research [27], the end-effector of the robotic device is wrapped around the cotton, and three linear servos on the end-effector rotate to achieve single flower pollination for the self-pollinated tomato. Many pollination robots use a nozzle to apply dry pollen, pollen mixed with nutrient solution [22], or plant hormones [6] in a ranged spray on the target, which can improve efficiency and tolerance of pollination. However, the non-specific pollination caused by spraying pollen over a range of flowers simultaneously adds to the challenge of flower and fruit thinning efforts.

5.4 Conclusion and Future Prospect

5.4.1 Conclusion

Aerial pollination for wind farms is already in practice, and plant protection UAVs have the potential to increase returns by expanding the ratio of parental rows planted in hybrid rice. Improving the aerial operating time of UAVs can enhance pollination efficiency by reducing the need for frequent recharging. The precision pollination method using UAVs described in [26] has demonstrated the feasibility of fully automatic flight to a flower using current hardware. In contrast, most pollination robots and bionic robots that use robotic arms, are still in the experimental stage. The cost of robotic arms and ROS platforms is still expensive for commercial applications. Considering the cost of tens of thousands of dollars for the pollination robot hardware, it would require the robot to pollinate an area of over 2 hectares per day to achieve a justifiable return on investment within a 2–3 year period. However, current pollination robots do not yet meet this commercial specification. Nevertheless, it is not impossible to simplify the robot arm to reduce costs and maintain the original accuracy. In the future, pollination robots may need to strike a balance and make trade-offs. While robotic bionic systems like the robobees described in [23] offer promising technology, they are currently cost-prohibitive and require strict operating conditions, making them unlikely to be commercially available in the near future. Further optimization of the robot in terms of vision and shooting accuracy will offer commercial solutions in the future. Specialized methods of pollination, such as ultrasonic pollination [28], have more inherent limitations for commercialization due to the need for fixed ultrasonic pollinators and the fact that not all crops are suited for mobile planted crop rows. Since it is unclear whether modulation frequency varies for different crops and how exposure time for single pollination relates to market price, it is difficult to estimate the cost of individual pollinations. Although pollination robots provide autonomous and precise control, thereby eliminating the human element from artificial pollination, the extent to which this technology can replace natural and manual pollination remains uncertain.

5.4.2 Future Prospect

Hardware is the only biggest challenge for all pollination devices. While these algorithms for image recognition and path planning still need to be improved, the biggest limitation of the whole pollination process is currently the hardware facility. A major challenge for aerial pollination is the power supply, as longer flight times require larger batteries or fuel tanks, and the weight of the items carried affects energy consumption. The accuracy of navigation is influenced by signal strength, particularly in agricultural areas situated in remote mountainous and hilly regions, far from urban centers. Due to the underdeveloped communication infrastructure and signal attenuation caused by dense crops, navigation can be significantly affected, especially during long-distance flights. While UAVs can assist in pollination, the process necessitates the installation of additional facilities, such as signal enhancement stations, on farmland, resulting in increased costs. In other hand, while binocular stereo cameras are preferred for accurate pollination in image recognition technology, their effectiveness can be hindered by the complex and changing natural environment, which may cause issues such as overexposure or underexposure, resulting in reduced image quality and accuracy of depth perception. Therefore cameras are required to be able to automatically adjust parameters such as exposure time, white balance, and autofocus. Moreover, the camera lens is susceptible to being obstructed by dust or rain, which can make it challenging to obtain clear images and consequently affect the accuracy of depth perception. For instance, in [9], a pollen solution was sprayed using two jet nozzles that were designed to clean the lens regularly. This was necessary because falling droplets could easily dirty the lens. During single plant pollination, such as with apple blossoms, UAVs are typically small in size. Nonetheless, strong outdoor winds can present challenges for UAVs in achieving precise pollination. In the future, it will be important to consider how to achieve accurate and consistent pollination under outdoor conditions.

References

1. Potts SG, Imperatriz-Fonseca V, Ngo HT et al (2016) Safeguarding pollinators and their values to human well-being. Nature 540(7632):220–229
2. Partap U, Ya T (2012) The human pollinators of fruit crops in Maoxian County, Sichuan, China. Mt Res Dev 32(2):176–186

3. Estravis-Barcala MC, Sáez A, Graziani MM et al (2021) Evaluating honey bee foraging behaviour and their impact on pollination success in a mixed almond orchard. Apidologie 52(4):860–872
4. Bao W, Peng G, Chen C (2017) Precise pollination technology of fragrant pear. Nong Cun Ke Ji 03:52–53
5. Li K, Huo Y, Liu Y et al (2022) Design of a lightweight robotic arm for kiwifruit pollination. Comput Electron Agric 198:107114
6. Shu Q, Liu Z, Zhang J et al (2015) The effects of different pollinators on pollination efficiency in kiwifruit. J Zhejiang Agric Sci 56(09):1416–1417. https://doi.org/10.16178/j.issn.0528-9017.20150920
7. Liu A, Zhang H, Liao C et al (2017) Effects of supplementary pollination by single-rotor agricultural unmanned aerial vehicle in hybrid rice seed production. Agric Sci Technol 18(3):543–552
8. Wang S, Lei X, Tang Y et al (2020) Spraying pollination technology for pear trees based on multi-rotor UAV. Jiangsu Agric Sci 48(23):210–214. https://doi.org/10.15889/j.issn.1002-1302.2020.23.043
9. Williams H, Nejati M, Hussein S et al (2020) Autonomous pollination of individual kiwifruit flowers: toward a robotic kiwifruit pollinator. J Field Robot 37(2):246–262
10. Lim JY, Ahn HS, Nejati M et al (2020) Deep neural network based real-time kiwi fruit flower detection in an orchard environment. arXiv preprint arXiv:2006.04343
11. Barnett J, Seabright M, Williams HAM et al (2017) Robotic pollination-targeting kiwifruit flowers for commercial application. In: PA17 International Tri-Conference for Precision Agriculture
12. Duke M, Barnett J, Bell J, et al (2017) Automated pollination of kiwifruit flowers. In: 7th Asian-Australasian Conference on Precision Agriculture, pp 1–5
13. Nejati M, Ahn HS, MacDonald B (2020) Design of a sensing module for a kiwifruit flower pollinator robot. arXiv preprint arXiv:2006.08045
14. Williams H, Bell J, Nejati M et al (2021) Evaluating the quality of kiwifruit pollinated with an autonomous robot. Field Robot 1:231:252
15. Yuan T, Zhang S, Sheng X et al (2016) An autonomous pollination robot for hormone treatment of tomato flower in greenhouse. In: 2016 3rd international conference on systems and informatics (ICSAI). IEEE, pp 108–113
16. Shi Q, Liu D, Mao H et al (2019) Study on assistant pollination of facility tomato by UAV. 2019 ASABE Annual International Meeting. American Society of Agricultural and Biological Engineers, p 1
17. Wu H (2014) Research of effection hybrid rice seed pollination with unmanned helicopter. Hunan Agricultural University
18. Wang Y, Bai R, Lu X et al (2022) Pollination parameter optimization and field verification of UAV-based pollination of 'Kuerle Xiangli.' Agronomy 12(10):2561
19. Li J, Zhou Z, Lan Y et al (2015) Distribution of canopy wind field produced by rotor unmanned aerial vehicle pollination operation. Trans Chin Soc Agric Eng 31(3):77–86
20. Jiyu L, Lan Y, Jianwei W et al (2017) Distribution law of rice pollen in the wind field of small UAV. Int J Agric Biol Eng 10(4):32–40
21. Yang X, Miyako E (2020) Soap bubble pollination. iScience 23(6):101188
22. Li G, Fu L, Gao C et al (2022) Multi-class detection of kiwifruit flower and its distribution identification in orchard based on YOLOv5l and Euclidean distance. Comput Electron Agric 201:107342
23. Wood R, Nagpal R, Wei GY (2013) Flight of the robobees. Sci Am 308(3):60–65
24. Chechetka SA, Yu Y, Tange M et al (2017) Materially engineered artificial pollinators. Chem 2(2):224–239
25. Sun Z, Guo X, Xu Y et al (2022) Image recognition of male oilseed rape (Brassica napus) plants based on convolutional neural network for UAAS navigation applications on supplementary pollination and aerial spraying. Agriculture 12(1):62

26. Hulens D, Van Ranst W, Cao Y et al (2022) Autonomous visual navigation for a flower pollination drone. Machines 10(5):364
27. Ohi N, Lassak K, Watson R et al (2018) Design of an autonomous precision pollination robot. In: 2018 IEEE/RSJ international conference on intelligent robots and systems (IROS). IEEE, pp 7711–7718
28. Shimizu H, Sato T (2018) Development of strawberry pollination system using ultrasonic radiation pressure. IFAC-PapersOnLine 51(17):57–60
29. Jiang J, Liu L, Fu R et al (2020) Non-horizontal binocular vision ranging method based on pixels. Opt Quant Electron 52:1–10
30. Mu X, He L, Heinemann P et al (2023) Mask R-CNN based apple flower detection and king flower identification for precision pollination. Smart Agric Technol 4:100151
31. Li G, Suo R, Zhao G et al (2022) Real-time detection of kiwifruit flower and bud simultaneously in orchard using YOLOv4 for robotic pollination. Comput Electron Agric 193:106641
32. Strader J, Nguyen J, Tatsch C et al (2019) Flower interaction subsystem for a precision pollination robot. In: 2019 IEEE/RSJ International Conference on Intelligent Robots and Systems (IROS). IEEE, pp 5534–5541

Chapter 6
An Investigation into Apple Tree Pruning and an Automatic Pruning Manipulator

Shang Shi, Zicheng Tian, Siyuan Jiang, and Zhao Zhang

Abstract Robots are the development direction for intelligent agriculture. The current robotics technology has penetrated all areas and aspects of agricultural production, especially in areas that require a large number of artificial fields. In the field of fruit production, the degree of mechanization of soil cultivation, fruit tree management, plant protection, irrigation, and harvesting is increasing. Apple tree pruning is completed by manual labor. In recent years, breakthroughs have been made in the field of agricultural robots in object capture, machine design, and machine kinematics analysis. This chapter reviews the autonomous technology progress in apple pruning.

Keywords Pruning manipulator · Kinematics · Dynamics · Structural Design

---

S. Shi · Z. Tian · S. Jiang · Z. Zhang
Key Laboratory of Smart Agriculture System Integration, Ministry of Education, Beijing 100083, China
e-mail: 3424341948@qq.com

Z. Tian
e-mail: 1909754858@qq.com

S. Jiang
e-mail: 3475458955@qq.com

S. Shi · Z. Tian · S. Jiang · Z. Zhang (✉)
Key Laboratory of Agricultural Information Acquisition Technology, Ministry of Agriculture and Rural Affairs, China Agricultural University, Beijing, China
e-mail: zhaozhangcau@cau.edu.cn

College of Information and Electrical Engineering, China Agricultural University, Beijing 100083, China

Z. Zhang and X. Wang (eds.), *Towards Unmanned Apple Orchard Production Cycle*, Smart Agriculture 6, https://doi.org/10.1007/978-981-99-6124-5_6

6.1 Introduction

Agricultural production plays an important role in the progress of civilization. The mechanization and automation of agriculture are aligned with the trend of the times and have ushered in tremendous growth. The economic benefits and completion efficiency brought by the automation and mechanization of agriculture have provided unlimited power for the development of agricultural production.

The fruit industry is an important pillar of China's agricultural economic structure and a key industry for achieving the organic unity of economic and ecological benefits. In the process of modern agricultural development, the fruit industry is often regarded as one of the important factors affecting national economic development. With the continuous optimization of the agricultural industrial structure and scientific level, the planting scale of the fruit tree industry is increasing daily. During the 13th Five Year Plan period, the green coverage rate of fruit trees exceeded 24%, and the economic benefits of the fruit industry reached 0.4 trillion yuan, accounting for 0.06% of the gross national product, which also shows that fruit tree planting has become the main economic source and industrial pillar of China's forest and fruit growing region [1]. Among them, apples, as the most commonly seen fruits in daily life, have a huge market demand, but China's level of mechanization in fruit tree pruning is still relatively low.

Fruit growers sometimes manually trim apple branches to increase the quality and quantity of their fruit, but this is time-consuming, labor-intensive, and tricky owing to the high degree of unpredictability in agricultural output. Therefore, more than 35% of the labor used in the production of apples is used in branch buildings. Traditional manual pruning has difficulty meeting the demands of rapid market development, so the study of agricultural robots has become a problem that needs to be given great attention, and agricultural robots not only improve production efficiency in agricultural production but also reduce production costs. Therefore, the mechanization and automation of pruning work hold great research value.

6.2 Capture of Objects

6.2.1 *Object Detection*

How to parse the information that computers can understand from the acquired images is the core problem of machine vision. Extracting information from an image has three levels: classification, detection, and segmentation. (Fig. 6.1).

While the detection focuses on a specific object target and obtains the category information and location information of the target, then the target object will be separated from the entire background, and descriptive information of the target will be determined. Then, the basic idea of object detection is positioning + recognition. Then, the machine needs to complete learning 'image classification' (compared to

Fig. 6.1 Three levels of image understanding

simple image recognition, which requires 'background') and 'judgment position' (outputting four coordinates to mark the anchor frame position).

There are two main types of deep learning in object detection: two-stage and one-stage methods. The former first identifies many candidate regions through a selective search region proposal (primarily using the sliding window method, the convolution operation results, and artificially trained classifiers to assess the likelihood that targets exist in this area). Then, it classifies them using convolutional neural networks, typically R-CNNs. Faster-RCNN replaces the original Selective Search algorithm with the RPN network, integrates it into the model, and stores the features extracted by the convolutional neural network in the video memory, which reduces the occupation of disk space, improves the training performance, and accelerates the training speed. (Fig. 6.2).

The two-stage object detection algorithm FPN (feature pyramid network) is proposed based on Faster R-CNN to solve the shortcomings of the original algorithm in dealing with multiscale changes and the poor effects on small target detection. Through continuous upsampling and cross-layer fusion, FPN ensures that the output features contain both underlying visual information and deep semantic information.

The feature pyramid network (as shown in Fig. 6.3) mainly includes three stages: bottom-up, top-down, and lateral connections. The bottom-up process is the process of feeding images into the backbone ConvNet to extract features. The size of the feature map output by the backbone is either constant or reduced by a factor of 2. Those layers whose output dimensions are unchanged are grouped into a stage, and then the output features of the last layer for each stage are extracted. Top-down is to upsample the feature map obtained at the top level and pass it down. The lateral connection mainly includes three steps: 1*1 convolution reduction dimension, fusing the obtained features with the feature map sampled from the previous layer, and after addition, a 3*3 convolution is required to obtain the feature output of the layer.

The latter directly transforms the target localization problem into regression problem processing, commonly YOLO and so on. Current methods based on candidate regions have advantages in accuracy and precision (Fig. 6.4).

The proposal of YOLO v3 mainly uses the new Backbone, a feature pyramid structure to detect targets at different scales, and the Leaky ReLU activation function.

The backbone used by YOLO v3 for feature extraction uses Darknet-53, which introduces a large number of residual structures compared to the Darknet-19 structure in v2 and uses a convolutional layer with a step size of 2 and a convolutional kernel size of 3×3 Conv2D instead of the pooling layer Maxpooling2D. The purpose of

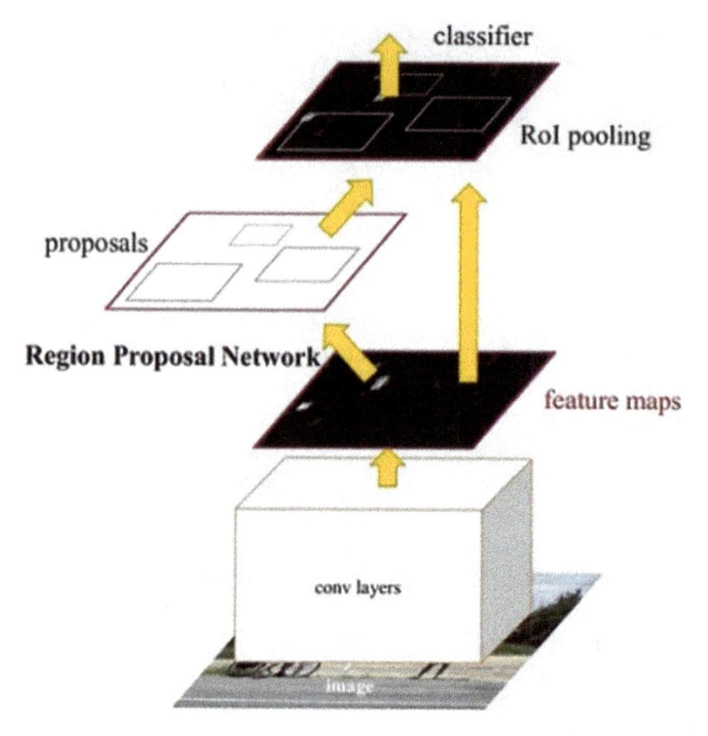

Fig. 6.2 The two stage type

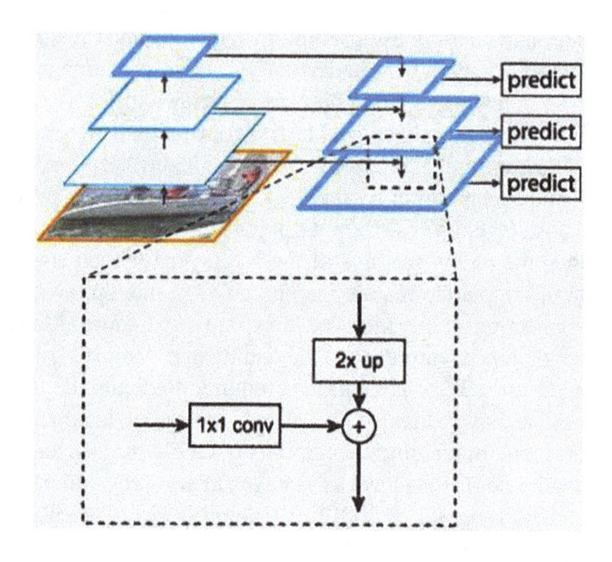

Fig. 6.3 Feature pyramid network

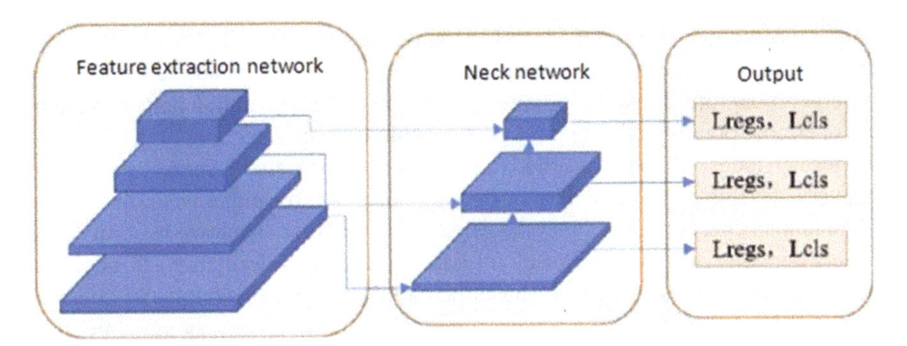

Fig. 6.4 YOLO series model structure

adding the residual structure is to increase the depth of the network, which is used to support the network to extract higher-level semantic features, and the structure of the residual can help us avoid the disappearance or explosion of gradients. The physical structure of the residuals, as reflected in the reverse gradient propagation, can cause the gradient to be transmitted to the network layer far ahead, weakening the chain reaction of reverse derivation.

6.2.2 3D Reconstruction

Generally, the problem of 3D reconstruction can be seen as the inverse problem of photo imaging, estimating the corresponding 3D shape based on 2D photo information. The representation of 3D data includes voxels, point clouds, implicit functions, and mesh representations. The current 3D object reconstruction is mainly based on the point cloud dataset ShapeNet, which utilizes similar geometric structures among objects of the same category. The reconstruction effect can be optimized if the object category is recognized and the corresponding general geometric structure is output. When point clouds are used to describe 3D models, they are intuitive and have unordered and irregular characteristics. Point clouds are defined as a massive set of points representing the target's surface characteristics, which can be simply understood as a collection of many feature points obtained by obtaining the spatial coordinates of each sampling point on the object's surface. Depending on the method of collecting the target, the point cloud may also contain information such as color and intensity, similar to the pixels of an image. The denser the point cloud is, the more details and attributes of the 3D object can be understood.

First, point cloud data are generally obtained by collecting point cloud data for objects without a basic 3D model to represent the 3D object. The methods for collecting point cloud data are mainly divided into active vision, passive vision, and active–passive vision combination. Active vision refers to using professional 3D scanning equipment to scan the target to obtain data. Common 3D scanning equipment includes LiDAR laser radar and TOF cameras. Passive vision refers to using cameras to capture the surrounding environment, such as natural light reflection, and obtaining point cloud data through algorithmic stereo matching of the obtained images or videos. The main methods include binocular stereo cameras and motion recovery structures. The structure of the binocular stereo camera is similar to that of the human eye, and it perceives the parallax information using two fixed-distance camera modules to obtain point cloud data with depth information. In addition, the motion recovery structure obtains point cloud data by algorithmically inferring the dataset obtained by a monocular camera. The motion recovery structure, also known as the SFM algorithm, is a typical algorithm applied to 3D reconstruction in the traditional computer vision field. Its principle is to use photos or video sources taken from multiple angles by a moving camera to match and infer the camera's pose parameters, calculate depth information through triangulation technology, and finally output point cloud data. For example, Zheng Lihua et al. used a Kinect 2.0 camera to collect multiangle information on fruit trees and then applied a series of algorithms to achieve a more accurate method of stitching the point cloud data of fruit trees [2].

Using a Kinect camera to capture images and then utilizing Microsoft's software development kit Kinect for Windows SDK and application programming interface (API), the RGB and depth information of the scene is fused to obtain point cloud information.

After obtaining the raw data, preprocessing is performed to remove noise points to obtain point cloud data containing only single fruit trees while reducing the amount of point cloud data to improve the accuracy and speed of subsequent point cloud processing. The noise removal idea is first to limit a range based on the obtained data and remove data outside the range, which can remove points far from the fruit tree's main body, using filtering methods. Then, the KD-Tree search algorithm is used to remove points that are not on the trunk or leaves but within the limited range. A normal aligned radial feature (NARF) key point search is used to normalize and align the fruit tree point cloud, and then the key points are extracted using fast point feature histograms (FPFHs). The mapping relationship between the two point clouds is estimated using feature values, and the mapping relationship is refined using the random sample consensus (RANSAC) algorithm to obtain the transformation parameters. Finally, the two point clouds are transformed into the same spatial coordinate system based on the transformation parameters, completing the initial registration of the point cloud. The ICP algorithm is then used for accurate registration (Fig. 6.5).

6.2.3 3D Positioning

The second step of automatic pruning is to determine the branches that need to be pruned and the location of the cut. First, the trunk and branches are separated, and then the length and diameter of each branch are calculated based on the previously

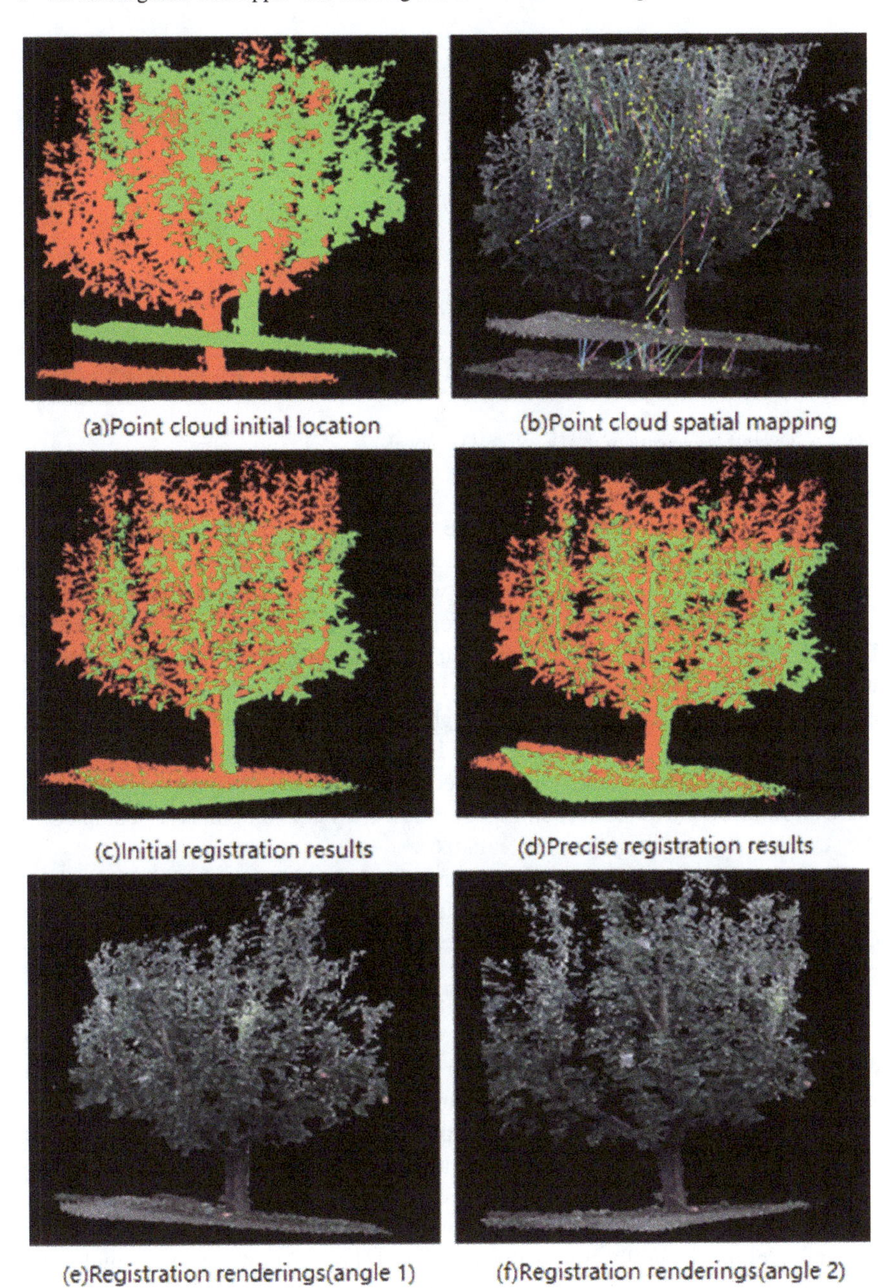

Fig. 6.5 Rendering

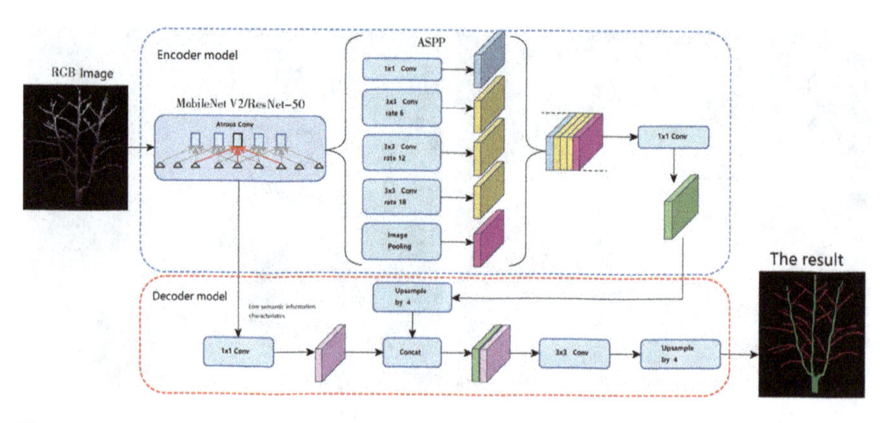

Fig. 6.6 Overall framework of the DeepLabV3+ model

extracted information. Then, whether the branch should be pruned and where to cut it is determined. The obvious difference between the trunk and branches is the direction. The trunk is generally upwards, while the branches are distributed on both sides of the trunk at a certain angle. Therefore, a semantic segmentation model, DeepLabV3+, which learns the geometric differences between fruit tree branches using a data-driven approach, can be used [3]. MobileNet V2 and ResNet 50 feature extraction networks are used to extract high and low semantic information features from fruit tree images. The high semantic information is further extracted using the atrous spatial pyramid pooling (ASPP) mechanism, and the obtained information is fused and then output to the decoding stage after a 1×1 convolution layer, completing the semantic segmentation process (Fig. 6.6).

Next, the length and diameter of the branches are calculated based on the data. According to Ma Baojian's method, the direction of a single pruning branch can be determined by calculating the maximum feature vector direction of the pruned point cloud [4]. Then, the point cloud is sliced along the pruning branch direction, and the center coordinates are calculated to fit the branch curve and obtain the length of the pruned branch. Similarly, the diameter is calculated by simulating a cylinder for each segment of the point cloud and taking the arithmetic mean of each segment; the more segments that are cut, the more accurate the results will be.

The principle of winter pruning for apple trees is mainly based on thinning and shortening [5, 6], removing branches that are too large or too dense, and shortening small and sparse branches. The previously calculated branch diameter can be used as the main basis for judging the size of the branch, and the length can be used to locate the cutting point, generally cutting the branch to 1/4 or 1/2 of its original length. The pruning standards for trees of different years and growth conditions may vary and can be flexibly set.

6.3 The Design of the Pruning Manipulator

Joint configuration (total degree of freedom and joint type) has a very important impact on the flexibility of the manipulator's movement, obstacle avoidance ability, and the final working status of the end effector (action scope and execution).

Considering the height of the crown of the apple tree, the growth of the branches, and the randomness of the growth direction of the branches, this chapter combines the design requirements of apple tree pruning and carries out the design of the size of each part of the pruning manipulator.

6.3.1 The Mechanism Design of the Manipulator

6.3.1.1 Positioning Mechanism of the Manipulator

The operation position of the end effector is mainly determined by three joints, which have the most impact on the spatial range of the manipulator work. According to the experience of mechanical engineering, the mechanical arm responsible for positioning operations must be considered. It is composed of the R-revolute joint and the P-prismatic joint. There are eight combinations. The following are the main four combinations:

1. Cartesian coordinate type (PPP): The three joints are all P-prismatic joints, and both are perpendicular to each other and finally form a square working area, as shown in Fig. 6.7.

   Such advantages are easy to control, large rigidity, and accurate positioning. Disadvantages: Smaller workspace.

2. Cylindrical coordinate type (RPP): A rotational movement, a vertical movement and a horizontal movement subcomposition, as shown in Fig. 6.8. In the end, a

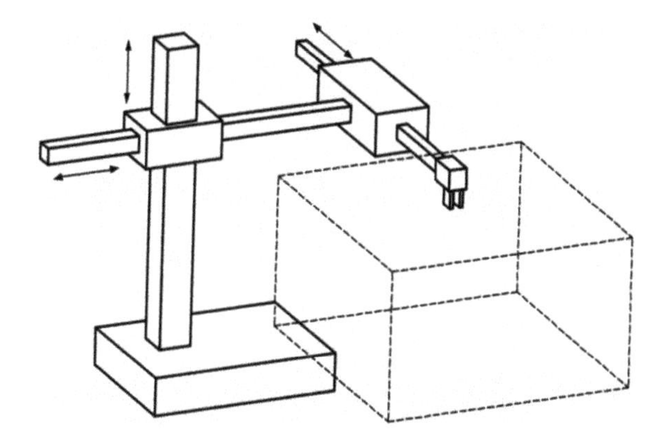

Fig. 6.7 Cartesian coordinate robot arm

column-shaped working area is formed, and the spatial positioning is relatively intuitive.

Advantages: The required space is small, the flexibility is high, and it offers a wide range of operations;

Disadvantages: Low positioning accuracy, limited load capacity and complicated control system.

3. Ball coordinate type: A rotary rotation, pitch movement, and horizontal movement subcomposition, as shown in Fig. 6.9. Finally, a spherical work area is formed.

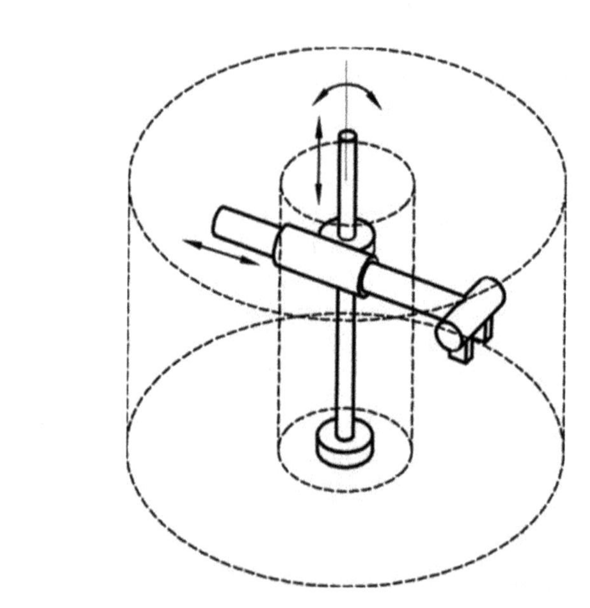

Fig. 6.8 Cylindrical coordinate manipulator

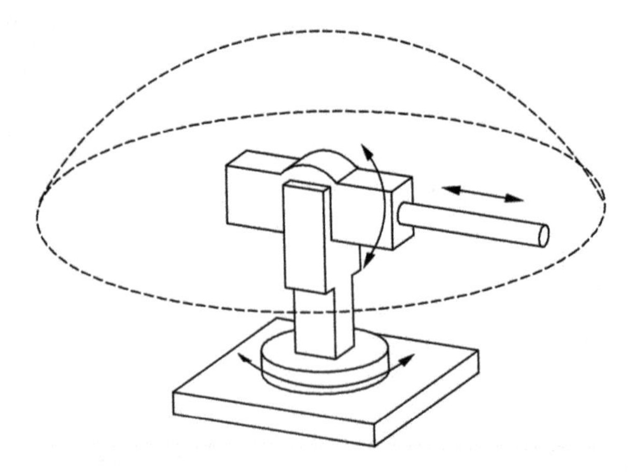

Fig. 6.9 Spherical coordinate arm

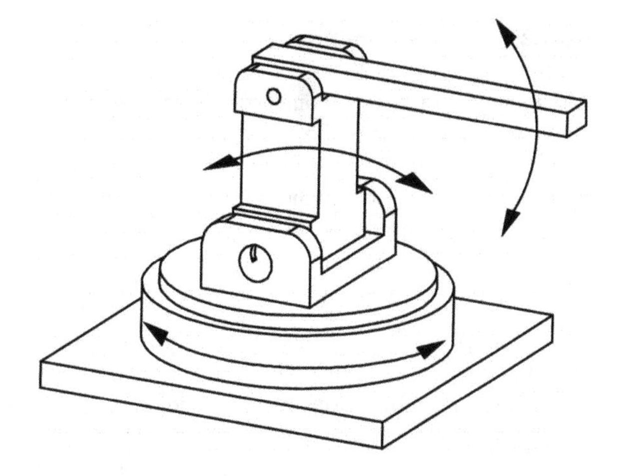

Fig. 6.10 Articulated robotic arm

Advantages: can be grasped on the ground operating objects with lower space positions; the structure is compact, and the working space is greater in the same parameter.

Disadvantages: The control accuracy is high, the position accuracy is low, and the load capacity is small.

4. Joint type: consisting of three rotating joints, as shown in Fig. 6.10.

Advantages: Compared to the ball coordinate model, the structure is more compact, the working space is larger, the flexibility is stronger, and the area is smaller.

Disadvantages: The control scheme is complicated, and the load capacity is low.

It can be seen that in the order of right-angle coordinates, cylindrical coordinates, ball coordinates, and joint types, the flexibility and working space of these four configurations are gradually increasing under the same parameters of the robotic arm. The load capacity gradually decreases.

Considering the pruning conditions of the apple tree itself and the biological characteristics of the apple tree:

1. The branches in the apple tree crown are dense. It is required that the robot has good flexibility to ensure that it can shuttle between branches
2. Pruning of apple branches, as long as the robot is required to bear its own weight, the load capacity is low.
3. The apple tree is high, the pole is high, and the size of the crown is large. It is required that the working space of the robot is large and can cover the crown.
4. Apple tree planting spacing is limited, and robotics should not occupy too much area to build smaller bases and walk in orchards.

Based on the advantages and disadvantages of the above four configurations and the characteristics of the apple tree, the positioning agency of the currently made apple tree trimming robot should choose the joint robotic arm.

6.3.1.2 The Position-Tuning Mechanism of the Manipulator

The manipulator's wrist connects the arm and the end-effector, is responsible for adjusting the position of the end-effector and has an independent degree of freedom. Select the degree of freedom reasonably according to work needs. There are three basic exercise methods on the wrist (as shown in Fig. 6.11). For three-dimensional space, 3 DOF-wrists are generally used. In this way, the flexibility of the end-effector can reach the peak efficiently, so that it can approach the apple tree in any direction.

In actual use, 3-DOF wrists can be divided into the following two structures:

1. Exchange wrist: The axis of the three joint axes of the exchange wrist remits on one point. The position relationship of the two adjacent axes can be divided into the orthogonal exchange type and oblique exchange type. This configuration complies with Pieper Guidelines (when one of the following conditions is met, a 6-DOF sports structure has the closed form of the kinematic inverse: 1. The axes of the three continuous rotation joints intersect at the same point, 2. However, the joint corners are limited by the mechanical structure, the flexibility is insufficient, and the first and third joint axes of the remittance will form a singular position.
2. Offset wrist: The joints of the partial wrist will intersect at two points. The gesture and position of the robot's end actuator are coupled, and the motion analysis is complicated. Generally, there is no analytical kinematic interpretation. However, the offset scale will reduce the limitation of the flexibility of the mechanical structure, which greatly improves the rotation range of each joint. Generally, the rotation range of the two joints can reach 360°, and the end-effector can

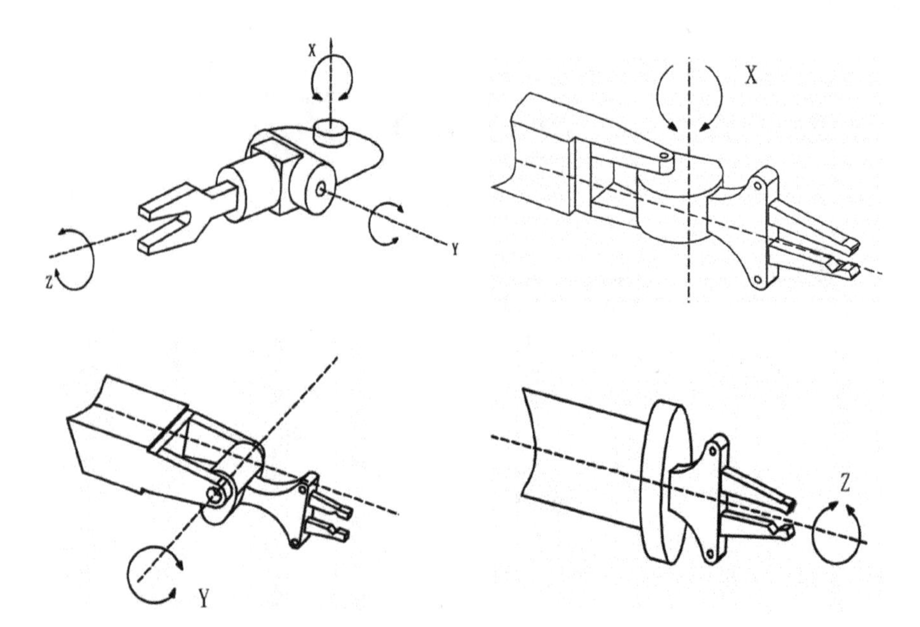

Fig. 6.11 Three basic movements of the wrist

be fine-tuned, which greatly increases the flexible obstacle avoidance of the manipulator.

When trimming, it is necessary to consider the collision between the manipulator and the nonoperating object. Therefore, after the robot positioning agency sends the wrist and the end actuator near the operation point, it should no longer move, relying on the end-effector's position between the branches only by the wrist. This has high requirements for the flexibility of the wrist, so we chose the offset wrist.

The 3-DOF offset wrist, under the combination of R-type joints (reincarnation joints) and B-type joints (swinging joints, which are conducive to structural compactness), also has different forms. Consider the impact of the construction conditions and choose the BRR-type wrist. The first wrist joint is swinging.

6.3.2 The Structural Design of the Manipulator

6.3.2.1 The Selection of the End-Effector

The terminal actuator is an important part of the apple tree-trimming robot. The joint motion of the robot sends the end-effector to the specified position after the end-effector completes the trimming operation. Based on the impact of the positioning accuracy of the robot's positioning of the robotic handling, the quality of the end-effector, and the ability of the wrist loading capacity, the apple tree trim of the robot's hand-end-effector should meet the following requirements:

1. For apple tree trimming operations, the most basic agronomic requirement is that the section of the branch separation should be smooth after trimming. If the cutting is not smooth, it will greatly impact the healing of the cutting port.
2. The spatial structure of the end actuator should be adapted to the highly nonstructured operating environment in the apple tree crown layer. The terminal actuator needs to reach into the relatively dense places of the apple branches for trimming, so there are certain requirements for the shape of the end actuator;
3. To meet the above requirements, the end actuator should also be as light as possible, have a simple structure, have good stability, and be low-ending.

The cutting–knife–type pruning device can smoothly separate the branches and meet the agronomy requirements of apple tree pruning (as shown in Fig. 6.12). Essence considering the requirements of trimming operations in the crown, the electrical-drive-type shear device is separated by the same diameter as the structure of the cylinder driver and hydraulic driver.

In summary, to make an apple pruning robot, this article selects the Yamaha-003 pruning on the market as the prototype of the end-effector.

Fig. 6.12 Branch shear separation section

6.3.2.2 The Structural Design of the Wrist and Arm of the Manipulator

The robotics of this article are mainly used for branch pruning operations. There is no additional load on the wrist except the end actuator. It only needs to drive the end actuator to adjust the position. From the dynamic analysis, we can see that the three wrist joints are not large, and the quality of the driving motor is small. At the same time, when pruning operations, the end actuator needs to be extended into the smaller space between the branches. The wrist adopts the BRR-type bias scheme. Maintenance is inconvenient, so a direct drive method is adopted. The limitation of the scope of the wrist joint rotation range of the wrist mechanical structure will affect the flexibility of the robotic hand. To improve the flexibility of robotic players, the wrist joints should not adopt the structural scheme where traditional industrial robots have built-in driving sources in the pole. When combining the wrist work, it must be shuttled in the limited space between the branches in the crown. This thesis designs a three-free gimbal structure with low manufacturing costs, convenient maintenance, high flexibility, and a small section size of the motor rod. The exploded view of the wrist is shown in Fig. 6.13.

Apple tree trimming robotics have no additional load except for weight, which is mainly used in agricultural production and should be considered from the aspects of cost and maintenance. From the dynamic simulation results, it can be seen that when the driver is directly driven, the driving moment required for the arm is not large; the elbow joint is consistent with the shoulder joint design scheme, and the driving torque generated by the drive motor with electromagnetic brakes. The key rotation directly drives the arm rotation. The exploded view of the arm is shown in Fig. 6.14.

6.3.2.3 The Structural Design of the Base of the Manipulator

The mass of each part of the robotic arm body will act on the base, and the base will bear a great force. Therefore, HT250, which has a high degree of stiffness and

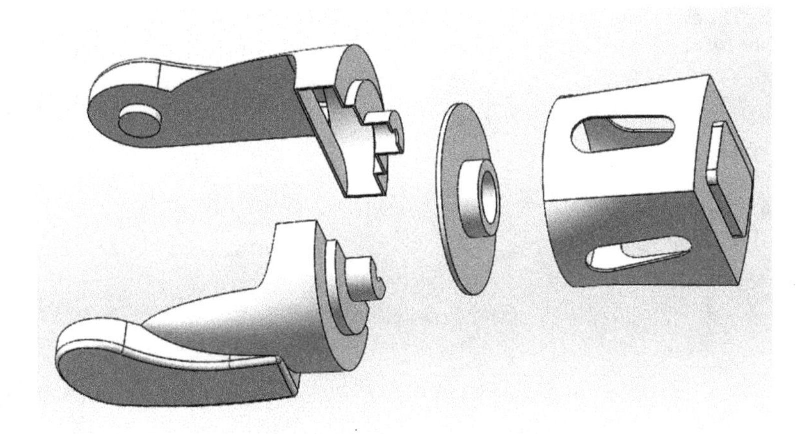

Fig. 6.13 The exploded view of the wrist

Fig. 6.14 The exploded view of the arm

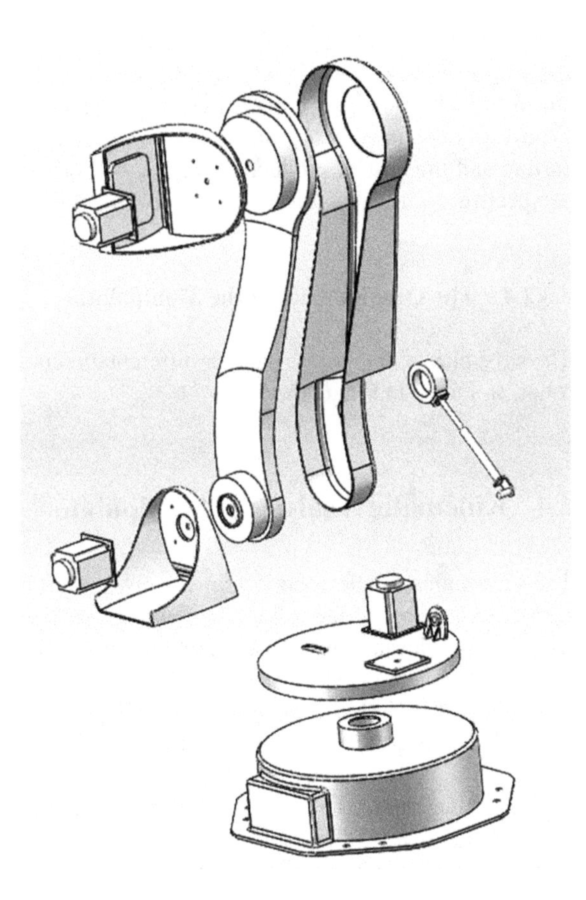

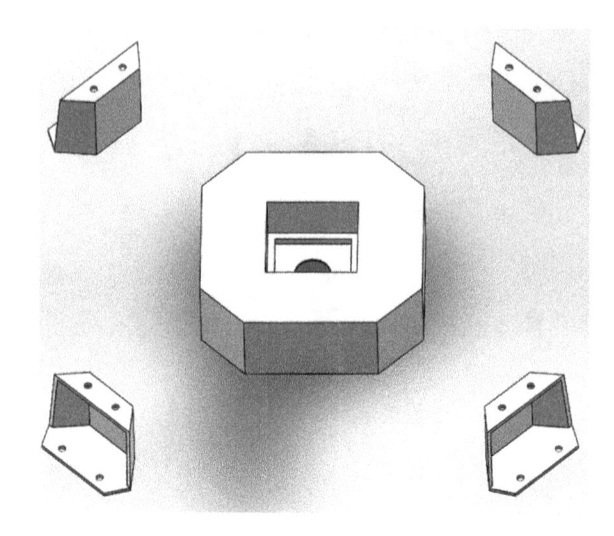

Fig. 6.15 The exploded view of the base

less impact vibration, is chosen as the basis material, and a permanent structure is added to each of the four corners to increase stability. In addition, the base is made to have an open interior layout, which lowers the amount of waste produced during casting and the cost of production. Figure 6.15 depicts the base from an exploded perspective.

6.3.2.4 The Omnipotence of the Manipulator

The wrist bias of the wrist is the four different directions of the robotic hand-pruning robot, as shown in Fig. 6.16.

6.4 Kinematic Analysis of Manipulator

The kinematics of the robotic arm mainly study the mutual motion relationship between the joints of the robotic arm and the spatial position between the base. It analyses the relevant transformation relationship of the joint position in the Cartesian coordinate system [7, 8].

Position description (Fig. 6.17): Establish the coordinate system An in three-dimensional space; then, any point P in space can be expressed as P(A) = [Px, Py, Pz].

Posture description (Fig. 6.18): If the arm structure of the robotic arm is simplified to a line segment with two endpoints, then when establishing the pose of the joint space, the state can be expressed by the coordinate system fixed on it, and a coordinate system B is fixed on the Q point. Then, the unit vectors Xb, Yb, and Zb in the direction

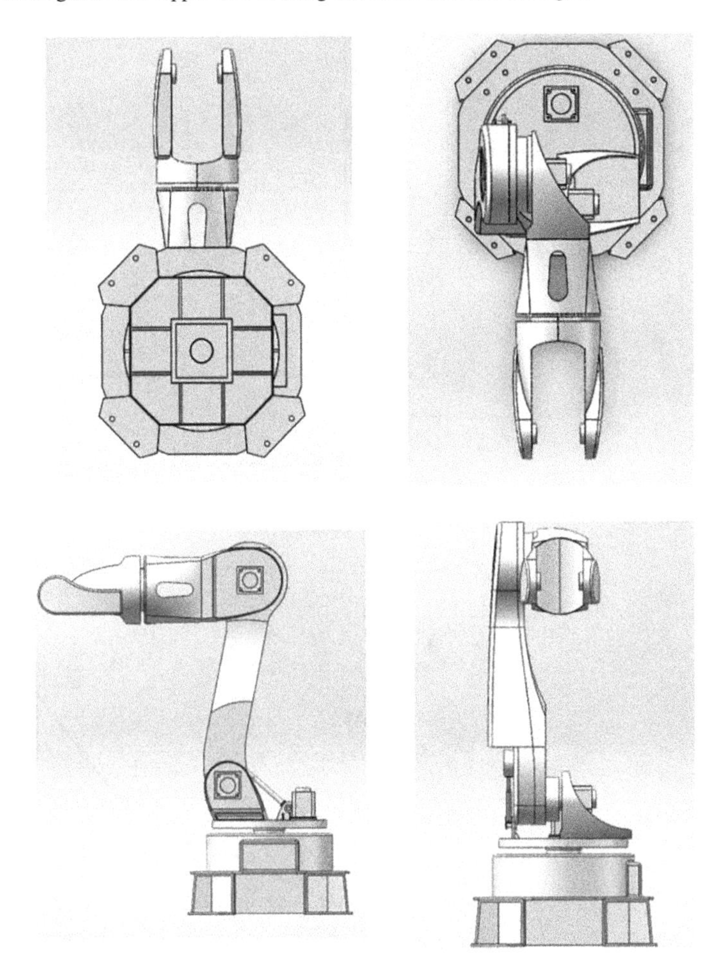

Fig. 6.16 The four views of the manipulator

of the main axis of the B coordinate system are represented in the A coordinate system: (A) Xb, (A) Yb, and (A) Zb. These three vectors are perpendicular to each other, and the following restrictions must be met: (A) Xb × (A) Yb = (A) Yb × (A) Zb = (A) Xb × (A) Zb = 0.

Spatial coordinate transformation: Spatial coordinate transformation includes translation transformation and rotation transformation, and the same point P in space can be represented by different coordinates in the two coordinate systems of (A) P AND (B) P. The translation transformation means that when the main axis direction of the two coordinate systems is the same and the origin coordinate is different, (A) P is obtained by translating along the direction of the line of the two origins to obtain the other coordinate (B) P, which is expressed as (A) P = (B) P + (A) PB. The rotation transformation means that the major axis direction is different but the coordinate origin is the same, and the description of point P in the A coordinate system obtains

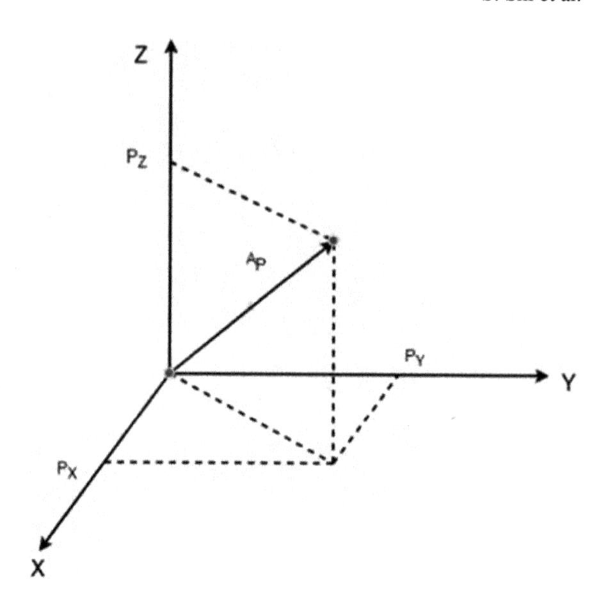

Fig. 6.17 Position description

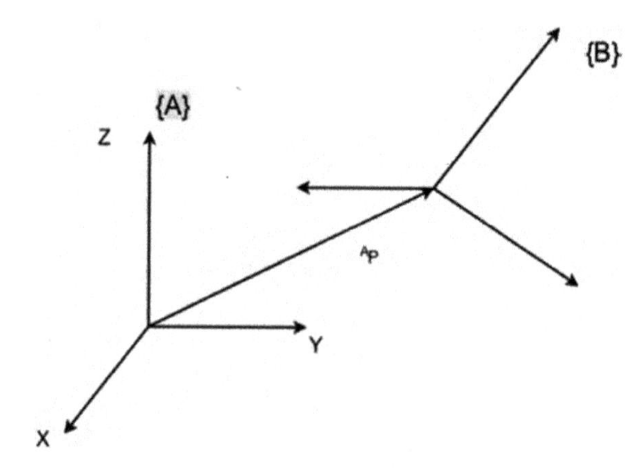

Fig. 6.18 Posture description

the position description in the B coordinate system through the rotation transformation, which is expressed as (A) P = (B)PRAB. (RAB is a 3 × 3 rotation matrix arranged in (A) Xb, (A) Yb, and (A) Zb).

6.4.1 D–H Parameter Method

The purpose was to obtain the relationship between the joint variables of the robotic arm and the position and attitude of the end effector. The posture relationship between the joints of the robotic arm is described by the homogeneous transformation matrix of 4 × 4, and the joint variables are obtained by solving the inverse kinematics.

The manipulator total transformation formula and the transformation matrix of the connecting rod coordinate system {n} and the connecting rod coordinate system {n + 1} are:

$$A_n = i - 1_{Tn} = Rot(Z, Q_1)Trans(0, 0, d_n)Trans(a_n, 0, 0)Rot(X, a_n)$$

$$ {}_{n}^{n-1}T = \begin{bmatrix} \cos\theta_n & -\sin\theta_n\cos\alpha_n & \sin\theta_n\cos\alpha_n & a_n\cos\theta_n \\ \sin\theta_n & \cos\theta_n\cos\alpha_n & -\sin\alpha_n\cos\theta_n & a_n\sin\theta_n \\ 0 & \sin\alpha_n & \cos\alpha_n & d_n \\ 0 & 0 & 0 & 1 \end{bmatrix}$$

Taking a three-degree-of-freedom series open-chain robotic arm as an example, the spatial transformation matrix of each joint of the robotic arm is:

$$A_1 = {}_{1}^{0}T = \begin{bmatrix} \cos\theta_1 & 0 & \sin\theta_1 & 0 \\ \sin\theta_1 & 0 & -\cos\theta_1 & 0 \\ 0 & 1 & 0 & 0 \\ 0 & 0 & 0 & 1 \end{bmatrix}$$

$$A_2 = {}_{2}^{1}T = \begin{bmatrix} \cos\theta_2 & \sin\theta_2 & 0 & L_2\cos\theta_2 \\ \sin\theta_2 & \cos\theta_2 & 0 & L_2\sin\theta_2 \\ 0 & 0 & 1 & 0 \\ 0 & 0 & 0 & 1 \end{bmatrix}$$

$$A_3 = {}_{3}^{2}T = \begin{bmatrix} \cos\theta_3 & -\sin\theta_3 & 0 & L_3\cos\theta_3 \\ \sin\theta_3 & \cos\theta_3 & 0 & L_3\sin\theta_3 \\ 0 & 0 & 1 & 0 \\ 0 & 0 & 0 & 1 \end{bmatrix}$$

From this, we can obtain: ${}_{3}^{0}T=A_1A_2A_3$. If you know the specific value of the joint angle, you can find the position and pose of the robotic arm. If the angle of each joint in the initial state is needed, the inverse kinematic analysis of the model is performed on the MATLAB instruction ikine to obtain the joint angle.

6.4.2 Path Planning

The trajectory planning of the robotic arm ensures that the position, velocity, and acceleration function curves of each corner joint variable are continuous in completing the work task. Through trajectory planning, the smooth movement of the robotic arm can be realized, vibration and shock can be reduced, and the accuracy of its motion trajectory can be ensured while avoiding additional energy consumption, reducing the wear of mechanical components, and improving the stability and accuracy of motion, thereby improving the efficiency of the operation and the effect of task completion.

Joint space trajectory planning uses the inverse kinematics method to convert the path points in Cartesian space into their corresponding joint angles and then uses the interpolation calculation method to calculate the interpolation points of each joint of the robotic arm so that the joints are synchronized and smoothly reach the target point. Trajectory planning now mostly uses cubic and quad-degree polynomial interpolation functions in polynomial interpolation calculation methods. The joint angle and angular velocity of the manipulator movement obtained by the cubic polynomial interpolation method are continuous. Its disadvantage is that there is no constraint on the angular acceleration of the joint, so its acceleration has a sudden change. The rate of change of acceleration is very large, which will lead to unstable manipulator movement and a certain impact on the mechanical system. The quinth-order polynomial interpolation method not only constrains the angle and angular velocity at the beginning and end of the path but also constrains the angular acceleration, so it can ensure a smooth and continuous joint angle, angular velocity, and angular acceleration.

6.5 Conclusion

This investigation is based on developing a robotic player that can meet the needs of apple trees as its core goal, mainly collecting the following research work.

1. On the basis of completing the extraction of the main information of the picture through three levels of classification, detection, and segmentation, the machine is required to complete the learning of the two tasks of 'image classification' and 'judgment position'. Through the use of YOLOV3 learning methods, YOLOV3 mainly includes the new backbone, namely, Darknet-53, the characteristic pyramid structure to detect the detection of different scale goals, and the Leaky RELU activation function. Deep camera dual-view shooting was used to obtain point cloud data and obtain a 3D model of fruit trees through a series of algorithm data filtering and standards. The branches are separated from the trunk through the semantic segmentation model, and the point cloud data are processed according to the set principles to determine the branch of the branches and the position of the cutting point.
2. Taking the apple tree as the pruning object, the actual work needs of the apple tree trimming have compiled the mechanical positioning mechanism and the tuning mechanism, and the mechanism parameters were given by the apple tree biological characteristics. Finally, a design incorporating the wrist bias was developed for the six-degree-of-freedom robotic arm.

3. Analysis of the motion science of the robotic arm, first based on the relevant conversion relationship in the Cartesian coordinate system, indicating the interrelationship between the joints between the robotic arm joints and the space between the base, including position description and attitude description. To complete the description of the coordinate relationship, the D–H parameter method represents the transformation matrix between the coordinate system and the space conversion matrix. Second, to complete the motion path planning, the five-time polynomial interpolation method improves the angle and angle speed of the path starting point and end point path to ensure the smooth and continuous consequences of the joint angle, angle speed, and angle acceleration and maintain the stability of the robotic arm movement.
4. Analysis of mechanical dynamics. Create a three-dimensional model dynamic simulation in SolidWorks, which provides a basis for structural design and drive selection.
5. The conclusions based on institutional design, pruning performance analysis, and dynamic analysis combined with the specific work of pruning operations require a structural design of the pruning robot, and the driving selection of each part is completed.

In summary, the wrist bias of the wrist in this article can meet the position of the end actuator in the apple tree with smaller space requirements in the entire crown layer. Pruning requirements. This article has a certain reference value in studying the research methods of apple tree pruning robots.

References

1. Wu B, Liu Y, He X, et al (2019) Current status and development tendency of fruit tree pruning equipment at home and abroad [J]. Hebei Fruit Tree 160(04):1–3+6. https://doi.org/10.19440/j.cnki.1006-9402.2019.04.001
2. Zheng L, Mai C, Liao W, Wen Y, Liu G (2016) 3D point cloud registration for apple tree based on Kinect camera [J]. J Agric Mach 47(5):9–14
3. Ma B, Yan J, Wang L, Jiang H (2022) Method for detection and skeleton of pruning branch of jujube tree based on semantic segmentation for dormant pruning [J]. J Agric Mach 53(8):313–319+442
4. Ma B (2022) Study on the key technologies of intelligent pruning for dwarf and dormant jujube tree [D]. Zhejiang University. https://doi.org/10.27461/d.cnki.gzidx.2022.001592
5. Guo X (2023) Shandong Yantai apple tree winter plastic pruning technology [J]. Spec Econ Anim Plants 26(1):110–112
6. Zhang Q (2022) Apple tree pruning principles and four seasons pruning technology thinking research [J]. Farmer's Staff 19:14–116
7. Sun Y, Zhang J, Li T, Jing Y, Pan X, Li M (2022) Kinematic analysis of apple picking robotic arm [J]. China Equip Eng 9:31–33
8. Ma Y, Zeng T, Jiang H (2023) Kinematics analysis and trajectory planning of manipulator based on MATLAB [J]. Packag Eng 2:188–190

Chapter 7
Apple Harvesting Robotics Review

Shahram Hamza Manzoor and Zhao Zhang

Abstract Apple is one of the most popular fruit in the world, and can be consumed in a number of ways, such as fresh eating, apple sauce, fresh, and juice. Apples are still manually harvested as the major approach. Due to increasing labor cost and shrinking labor pool, apple harvest is the challenge faced by growers. With technology progress, apple harvest robots are gradually developed, which consists of apple detection, localization, picking and placement. This chapter mainly focuses on describing the technology progress on apple harvest.

Keywords Apple · Harvest · Robotics · Detection

7.1 Introduction

The world's apple production is expected to reach 78.8 million metric tons in the 2022/23 according to the report published by United States Department of Agriculture Foreign Agricultural Service, USDA. Apples are grown in many regions around the world, with some countries standing out as major producers. China is currently the world's largest producer of apples, accounting for over 48% of global production. Other significant producers include the United States, Turkey, Poland, Italy, and India. These countries have favorable climates and soil conditions for apple cultivation, and many have a long history of apple growing and exporting. Apples are a valuable crop for many of these countries, both for domestic consumption and

S. H. Manzoor · Z. Zhang
Key Laboratory of Smart Agriculture System Integration, Ministry of Education, Beijing 100083, China
e-mail: shahramhamza786@uaar.edu.pk

S. H. Manzoor · Z. Zhang (✉)
Key Lab of Agricultural Information Acquisition Technology, Ministry of Agriculture and Rural Affairs, China Agricultural University, Beijing, China
e-mail: zhaozhangcau@cau.edu.cn

College of Information and Electrical Engineering, China Agricultural University, Beijing 100083, China

Z. Zhang and X. Wang (eds.), *Towards Unmanned Apple Orchard Production Cycle*, Smart Agriculture 6, https://doi.org/10.1007/978-981-99-6124-5_7

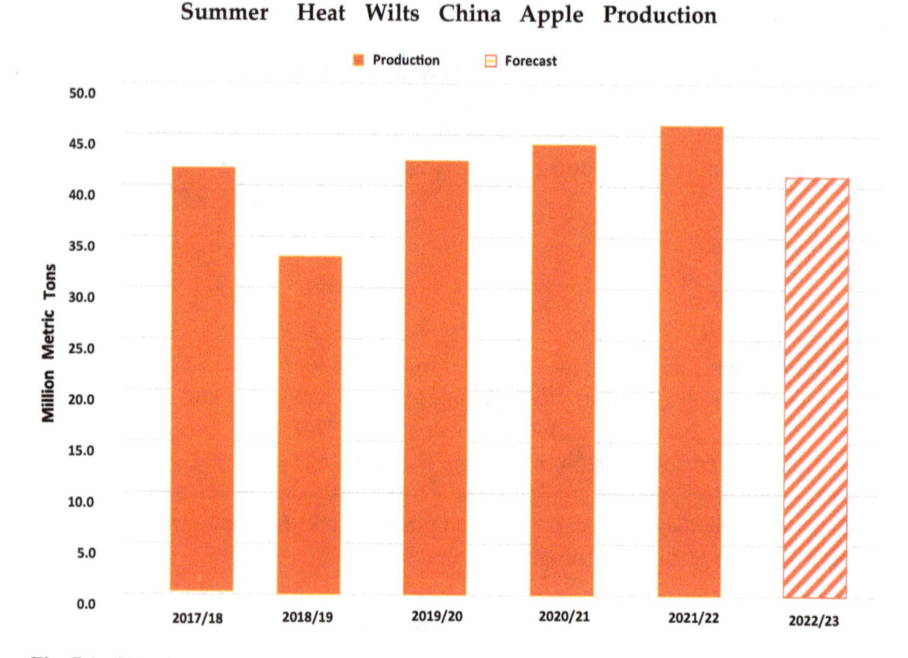

Fig. 7.1 China's apples production statistics. *Source* United States Department of Agriculture Foreign Agricultural Service, USDA

for export to other regions. As global demand for apples continues to grow, these major producing countries are likely to remain important players in the global apple market. In terms of apple production, China is the clear leader with 44,447,793 metric tons produced annually. The United States follows with much lower but still significant 4,649,323 metric tons. Poland, Turkey, India, Iran, Italy, Russia, France, and Chile round out the top ten producers, with Poland producing 3,604,271 metric tons, Turkey producing 2,925,828 metric tons, India producing 2,872,000 metric tons, Iran producing 2,799,197 metric tons, Italy producing 2,455,616 metric tons, Russia producing 1,843,544 metric tons, France producing 1,819,762 metric tons, and Chile producing 1,759,421 metric tons (Figs. 7.1 and 7.2).

7.1.1 *Challenges Facing the Global Apple Industry: Issues in Production and Supply*

There are several issues that the apple industry is facing globally. Some of the most significant ones are:

Climate Change: Changes in weather patterns and extreme weather events, such as droughts and floods, have affected apple production in many regions. Climate

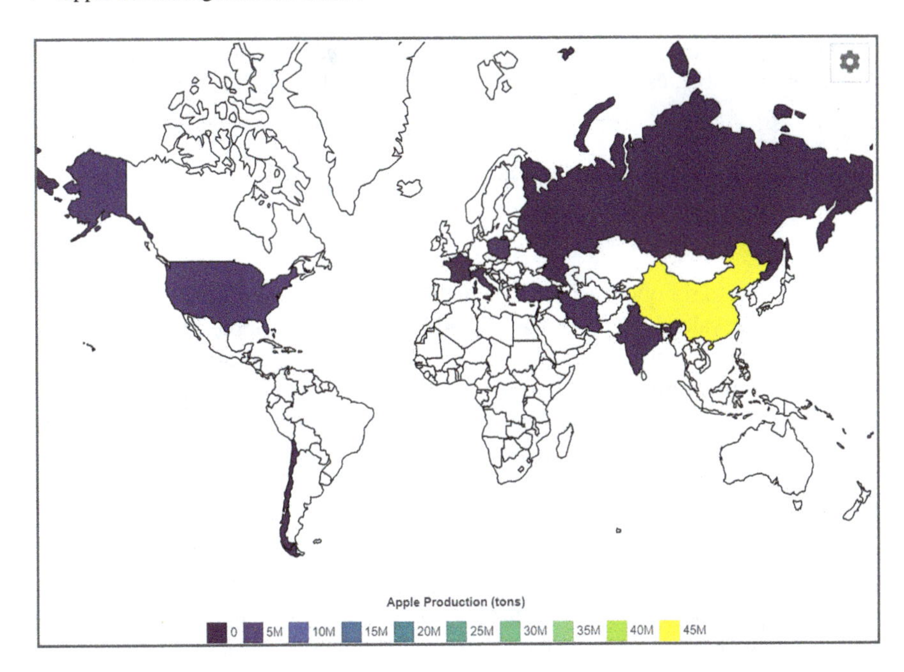

Fig. 7.2 Global map apples producing countries. *Source* World population review

change is also causing apple trees to bloom earlier or later, leading to challenges in harvesting.

Pests and Diseases: Fire blight, apple scab, and codling moth, for example, can have a significant influence on apple productivity. These challenges frequently necessitate the use of pesticides, which can cause environmental issues as well as decrease apple quality.

Labor Shortage: Apple harvesting is a laborious procedure, and many countries are battling a labor shortage. This can result in higher labor expenses and harvesting delays, which can influence apple quality.

Market Competence: Other fruit producers, such as oranges and grapes, compete with the apple sector. Furthermore, competition from other snack foods and beverages, such as energy bars and soft drinks, is expanding.

Transportation and Logistics: Apples are edible items, therefore transportation and logistics are essential to ensure that they reach in good condition. However, concerns like delays and improper storage can cause rotting and reduce apple quality.

Consumer Preferences: Because consumer preferences are continuously changing, the Apple industry must adapt. Organic and locally grown apples, for example, are in high demand, as are novel apple kinds and flavors.

Cost of Production: Due to factors such as labor expenses, equipment and machinery expenditures, and the cost of pesticides and fertilizers, the cost of producing apples can be considerable. This can make it difficult for growers to remain profitable, especially in areas with high land and labor expenses.

7.1.2 Introduction of Robotics in the Apples Production Industry

The application of robotics as a solution in the apple producing business is becoming increasingly prevalent. Many of the operations involved in apple harvesting, from selective picking to sorting and packing, are being automated using robotics technology. One of the primary benefits of adopting robotics in apple production is their capacity to work nonstop without interruption. Robots, unlike human laborers, do not require rest intervals, which increases the speed and efficiency of the harvesting process. In addition, because robots can work in a range of weather situations, growers may harvest their crops more quickly and efficiently, avoiding the need for extra labor. Another big benefit of adopting robotics in Apple production is that it can assist prevent worker injuries. Harvesting apples is a physically demanding operation, and robots can do it with precision and regularity while putting workers' health and safety at risk. This means that farmers can minimize the costs of workers' compensation claims while also reducing downtime due to injuries. Robotic apple harvesters may also perform selective picking, which means they can identify and pick only the ripest and healthiest apples, leaving behind others that are not yet ready for harvest. This can result in higher-quality apples and potentially less waste generated during the harvesting process. Growers can boost their revenue while lowering their environmental effects by eliminating waste. The use of automation in apple production may help draw younger people to the sector. With the rise of automation and technology, the agriculture sector may become increasingly enticing to young individuals interested in STEM-related occupations. Long-term labor shortages can be addressed by encouraging more people to enter the sector. Robotics technology in apple production can assist overcome labor shortages by providing a dependable, cost-effective, and safe alternative to traditional labor-intensive harvesting methods. The utilization of robotics in apple production has the potential to enhance operational efficiency and increase output, thereby leading to the creation of superior quality products. As technological advancements continue to occur and costs decrease, it is anticipated that the implementation of robotics will become more widespread in the apple production sector.

7.1.3 *Revolutionizing Apple Harvesting: The Advantages of Opting for Robotics.*

The apple industry holds significant importance in the global agricultural economy. Due to the scarcity of agricultural laborers in several nations, there is a rising inclination towards the incorporation of robotic technology to mitigate labor shortages in apple cultivation. In recent times, there has been a rapid advancement in robotic technology, which has found diverse applications in apple harvesting, ranging from pruning to picking. The purpose of robotic apple harvesters is to autonomously traverse through orchards and efficiently gather apples while minimizing the need for human intervention. The machines are equipped with cameras and sensors that facilitate the identification of mature fruit and enable the determination of the most efficient harvesting method. Upon detecting an apple, the robot proceeds to extract the fruit from the tree with precision and subsequently deposits it into a receptacle, utilizing either a gripper or suction cup mechanism. A significant benefit of utilizing robotic technology for apple harvesting is its ability to operate continuously, thereby reducing the requirement for supplementary manual labor. These devices are capable of functioning in diverse weather conditions and during nighttime, thereby augmenting their efficiency and productivity. In addition, automated apple harvesting machines exhibit the ability to engage in discerning picking practices, whereby they can discern and collect solely the most mature and robust apples, while disregarding the underdeveloped ones. This phenomenon may lead to the production of apples with superior quality and a possible reduction in the amount of waste generated while harvesting.

The development of robotic pruning technologies is underway to facilitate the upkeep and contouring of apple trees. The aforementioned devices employ sensors and cameras to examine the tree's anatomy and identify the optimal branches for pruning or removal.

Once the system has recognized the branches that need to be clipped, it makes the cuts with a robotic arm equipped with pruning shears. By removing dead or damaged branches and fostering new growth, this technique can assist enhance tree health and yields. The application of robotic technology in apple production is fraught with difficulties. The expense of purchasing and maintaining the machinery is one of the major challenges. Furthermore, because the technology is still relatively new, some growers may be unwilling to invest in it until it has been validated. Nevertheless, as technology advances and the agricultural labor deficit persists, robotic apple harvesting and pruning systems are anticipated to become more prevalent in the industry.

Ultimately, the use of robotic technology in apple harvesting and pruning has the potential to transform the business by providing a dependable and cost-effective alternative to existing labor-intensive harvesting methods. These machines have the potential to boost efficiency, lower labor costs, and result in higher-quality apples. As technology advances, we should expect to see even more creative solutions to the difficulties confronting the apple business.

7.1.4 Major Companies Leading the Development of Apple Harvesting Robotics and Technologies

Several prominent corporations are involved in the development and manufacture of apple harvesting robotics and technology. Among these businesses are:

Abundant Robotics: Abundant Robotics, headquartered in California, has created an apple-harvesting robot. The robot gently removes apples from trees and places them in a bin using a vacuum-based method. The robot, which can pick up to 10,000 apples per day, is currently in operation in various orchards around the United States (Fig. 7.3).

FF Robotics: FF Robotics is a Norwegian firm that has created an apple harvesting robot. To identify and harvest ripe apples, the robot employs a combination of cameras, sensors, and algorithms. The robot can gather up to 300 apples each hour and is presently in operation in various European orchards.

Dogtooth Technologies: Dogtooth Technologies is a Canadian firm that has developed an apple harvesting robotic arm. To recognize and pluck ripe apples, the robotic arm employs machine vision and artificial intelligence. The arm is coupled to a vehicle that can traverse the orchard, plucking apples as it goes. This company is now developing a second-generation robotic arm that will be more efficient and effective than the first (Fig. 7.4).

Energid Technologies: Energid Technologies is a Massachusetts-based firm that has created an apple harvesting robotic arm. To locate and harvest ripe apples, the robotic arm employs modern sensors and algorithms. The arm is positioned on a movable base that can traverse the orchard, collecting apples as it goes. Currently, the company is aiming to improve the efficiency and speed of their apple harvesting robot (Fig. 7.5).

Harvest CROO Robotics: Based in Florida, has built a strawberry picking robot and is working on an apple picking robot. The apple picking robot will recognize and harvest ripe apples using modern sensors and machine vision. The robot

(a) (b)

Fig. 7.3 Abundant Robotics apple picking machine prototype (**a**), and (**b**)

Fig. 7.4 Dogtooth Technologies harvesting machine

Fig. 7.5 Energid Technologies apple harvesting machine

is still in development, but the business anticipates having a working prototype in the near future (Fig. 7.6).

These are only a handful of the prominent firms active in the research and development of apple harvesting robotics and technologies. As the demand for more efficient and cost-effective apple harvesting methods grows, it is expected that more companies will enter this industry in the coming years.

Fig. 7.6 Harvest CROO Robotics harvesting machine

7.1.5 Traditional Methods of Apple Harvesting: Issues and Challenges

Apple picking is usually done by hand, with people scaling ladders to gather the fruit from the trees. This technique is called handpicking, and it has been used for many years. Handpicking, on the other hand, can be a time-consuming and labor-intensive process because workers must wander throughout the orchard and climb up and down ladders to collect the fruit. This can make the procedure inefficient, particularly in large-scale commercial orchards where time and cost are crucial. An-other difficulty with handpicking is the possibility of worker injury. Climbing ladders to gather fruit can be hazardous, and employees may fall and injure themselves. This can be unsafe not only for the employees but also expensive for the orchard owners, who may be held accountable for any accidents experienced by the workers.

Handpicking can potentially cause fruit damage if it is not done carefully. This can diminish crop quality and value, resulting in financial loss for orchard owners. Furthermore, handpicking may not be able to keep up with the enormous demand for apples, particularly during peak harvest season, resulting in harvest delays and decreased efficiency.

Modern robotics and equipment have been created to overcome these concerns and improve the efficiency and safety of apple harvesting. These devices can shake the trees to extract the fruit, eliminating the need for physical labor, or they can employ drones outfitted with cameras and sensors to detect ripe fruit on the trees. Robotics and grippers can also be utilized to remove the fruit gently from the trees, reducing the risk of fruit damage and the necessity for manual labor (Figs. 7.7 and 7.8).

Fig. 7.7 Handpicking of apples

Fig. 7.8 Apple harvesting using ladders

7.1.6 *Modern Robotics and Machines Improving Apple Harvesting*

Currently, there exists a wide range of mechanical apple harvesting devices that employ shaking mechanisms to dislodge fruit from trees, thereby mitigating the necessity of human intervention.

The OTR Grape Harvester is a specialized piece of equipment that has been developed for the purpose of efficiently gathering grapes and other types of fruit from vineyards. The mechanism employs a sequence of pliant rubber appendages that apply a mild agitation to the vines, thereby dislodging the produce, which subsequently descends onto a conveyance apparatus and is transported to a repository container.

The Korvan 3016 Harvester is a piece of agricultural machinery designed for the purpose of harvesting crops. The aforementioned apparatus is a harvester that is capable of self-propulsion and is designed to be towed behind another vehicle. It employs a set of shaker rods to effectively dislodge apples from their respective trees. The fruit descends onto a sequence of conveyors that transport it to a receptacle for gathering (Fig. 7.9).

The KOKAN 5000 Apple Harvester is a self-propelled, tow-behind machine that employs a shaking mechanism to dislodge apples from the trees. The fruit is conveyed through a series of conveyors and subsequently deposited into a collection bin (Fig. 7.10).

Fig. 7.9 Korvan 3016 Harvester

Fig. 7.10 .KOKAN 5000 Apple Harvester

Darwin the Cherry and Apple Harvester is a specialized piece of equipment that has been developed for the purpose of harvesting cherries and apples. The device employs a sequence of pliable, rubberized appendages that delicately extract the produce from the arboreal structure and deposit it into designated receptacles (Fig. 7.11).

Oxbo 7440 the Blueberry Harvester is a specialized piece of equipment that has been purposefully engineered to efficiently gather blueberries from their plants. The process involves the utilization of a set of tines that systematically traverse through the foliage to extract the produce, subsequently transferring it to a designated receptacle for aggregation (Fig. 7.12).

The FMR Smart 120 is a mechanized harvester that is capable of self-propulsion and can be towed behind a vehicle. It employs a shaking mechanism to effectively dislodge apples from their branches. The fruit descends onto a sequence of conveyors which facilitate its transportation to a receptacle for gathering. The apparatus is furnished with an optical sensor which enables it to perceive the degree of maturity of the fruit and regulate its vibration intensity correspondingly.

The subject of discussion pertains to the geographical location known as New Holland.

The Braud 9090X Olive Harvester is a specialized piece of equipment intended for the purpose of gathering olives, although it has the capability to be utilized for the collection of apples and other types of fruit as well. The process involves the utilization of a sequence of shaker rods to effectively dislodge the fruit from

Fig. 7.11 Darwin Cherry and Apple Harvester

Fig. 7.12 Oxbo 7440 Blueberry Harvester

Fig. 7.13 New Holland Braud 9090X Olive Harvester

the trees, which is then subsequently transported to a designated collection bin. The apparatus is outfitted with a sorting mechanism that segregates the fruit from extraneous matter such as foliage and other detritus (Fig. 7.13).

7.1.6.1 Machine Learning and Deep Learning for Apple Detection and Localization

Machine learning algorithms can be trained to identify ripe fruit on the trees and adjust the harvesting process accordingly, improving efficiency and reducing waste.

[1] This research uses RGB-D cameras with radiometric capabilities and multi-modal deep learning techniques to detect Fuji apples. They propose a method using CNNs and LSTM networks to process the multi-modal data. The dataset was used to train and evaluate the proposed approach, and it achieved high accuracy in detecting Fuji apples. The study demonstrated the potential of multi-modal deep learning techniques for fruit detection applications (Fig. 7.14).

[2] This research uses an RGB-D camera and machine learning algorithms to detect red and bicolored apples on trees. They propose a method that uses a random forest classifier and color thresholding technique with depth information to remove false detections. The study includes the development of a dataset and achieved high accuracy in detecting apples on trees. The results demonstrate the potential of using RGB-D cameras and machine learning for apple detection (Fig. 7.15).

Fig. 7.14 A view of the acquisition equipment showing the Kinect v2 sensors mounted on the mobile platform [1]

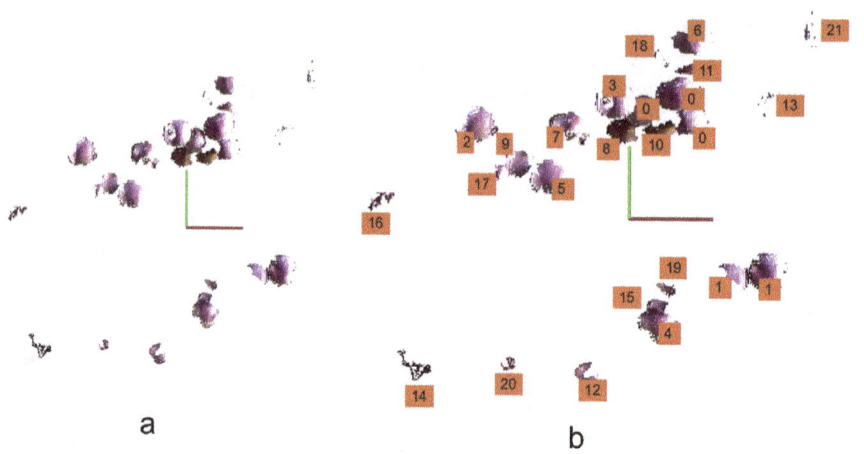

Fig. 7.15 Description of multiple steps in the distance and color filtered point cloud's Euclidean clustering segmentation: **a** identified clusters; **b** labeled clusters [2]

Fig. 7.16 Color distribution of related channels of a detected apple by deep learning. **a** Original image of detected apples. Channel distribution map of apple in red box was draw. **b** Channel distribution map of HSV, RGB, and LAB [4] (for interpretation of the references to color in this figure legend, the reader is referred to the web version of this article)

[3] This research develops the KFuji RGB-DS database for detecting Fuji apples using multi-modal images that include color, depth, and range-corrected IR data. They propose a method for integrating the data using CNNs and LSTMs and achieved high accuracy in detecting apples. The study demonstrates the potential of multi-modal imaging for fruit detection applications. [4] This research study proposes an improved binocular localization method for apple based on fruit detection using deep learning. The method involves using a deep learning model to detect apples in images captured by a binocular camera system, which are then used to calculate the location and orientation of the fruit. The proposed method improves upon previous approaches by reducing errors caused by occlusions and variations in lighting conditions. The method has the potential to increase the accuracy and efficiency of fruit harvesting and reduce labor costs (Fig. 7.16).

[5] Researchers used Deep Learning and YOLO-V4 to detect and count ripe Dezful native oranges in an orchard. The method achieved a precision of 91.23%, recall of 92.8%, F1-score of 92%, and mAP of 90.8%. The method can provide an effective solution for detecting and estimating yield in an orange orchard. [6] A computer vision algorithm was proposed to detect cutting points on grape clusters in vineyards. The method achieved an average recognition accuracy of 88.33% and a success rate of 81.66% in visually detecting cutting points on the peduncles. The method could be used by harvesting robots for efficient use in a complex environment. [7] A pruned YOLO V5s model was developed to detect apple fruitlets before thinning. The pruned model achieved a recall of 87.6%, precision of 95.8%, F1-score of 91.5%, and false detection rate of 4.2%. The average detection time was 8 ms per image, and the model size was 1.4 MB. [8] Machine vision was used to detect apples with incompletely red skins and various colors. The proposed method achieved an

average recall, precision, and F1-score of 89.80%, 95.12%, and 92.38% respectively. The method performed better than pedestrian detection and faster region-based convolutional neural network but was not robust to noise and had a slightly longer processing times. [9] This research study proposes a frustum-based point-cloud-processing approach for detecting and localizing occluded apples in robotic harvesting. The method involves processing 3D point cloud data obtained from a 3D camera mounted on a robotic arm to generate frustums, which are then classified as containing either apples or background objects. The approach is designed to handle occlusions, which can occur when apples are partially hidden behind leaves or other objects, and accurately detect and localize apples for efficient harvesting as shown in the Fig. 7.17.

[10] The study presented a novel approach for apple image segmentation in apple-picking robots, utilizing a multi-feature patch-based technique and the gray-centered RGB color space. The proposed method was found to be efficient and precise in identifying apple targets. The approach yielded an average precision rate of 98.79%, a retrieval rate of 99.91%, an F1 score of 99.35%, a type I error rate of 0.04%, and a type II error rate of 1.18%. [11] A new approach was suggested for detecting apples within the orchard setting, specifically for the purpose of facilitating the operation of a harvesting robot. The Shuffle-net v2-YOLOX approach was employed, yielding a precision of 95.62%, recall of 93.75%, and an F1-score of 0.95. The method exhibited a detection speed of 65 frames per second and outperformed other advanced lightweight networks in terms of both detection accuracy and speed. [12] The proposed apple recognition method based on improved YOLOv4 achieved high

Fig. 7.17 The apple fruits under different illuminations and occlusions [9]

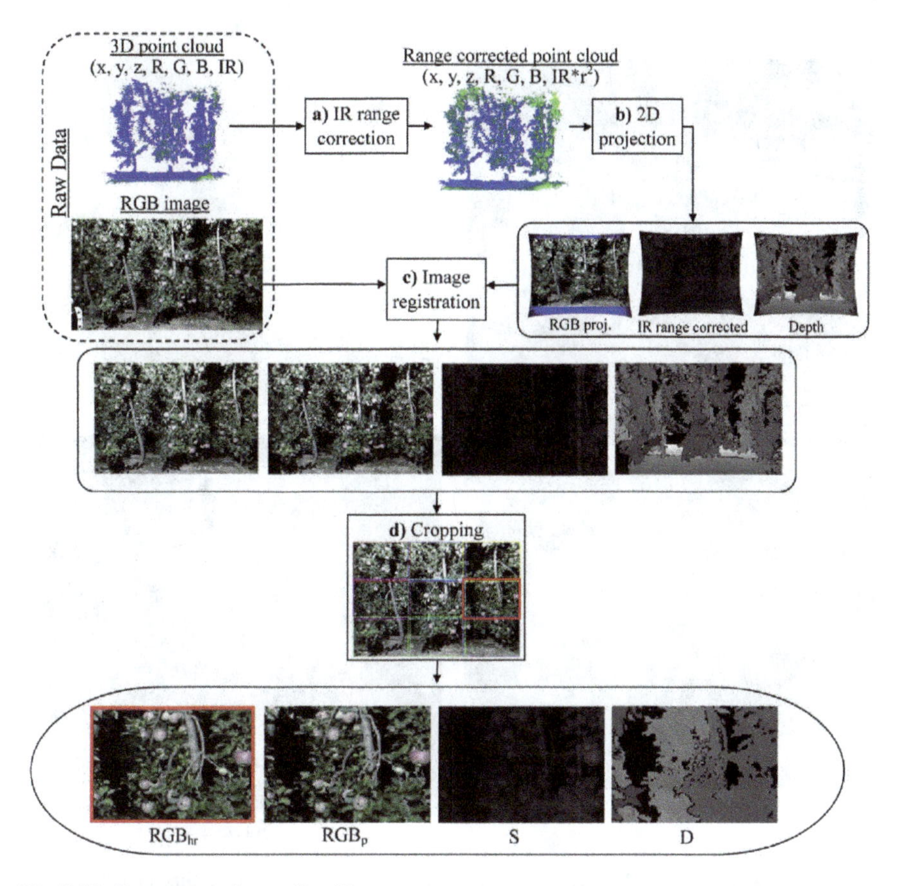

Fig. 7.18 Data preparation outline 22

accuracy (93.42% precision, 87.64% recall, and 0.9035 F1-score), while reducing storage memory by 87.8% and increasing recognition speed by 43%. The model is suitable for apple recognition scenarios such as fruit sorting and inventory management. The proposed algorithm uses several techniques, including feature fusion and attention mechanisms, and a novel training strategy to enhance accuracy and reduce training time. The method is efficient, accurate, and suitable for real-world applications (Figs. 7.18, 7.19, and 7.20).

7.1.6.2 Robotics: Robotic Arms and Grippers for Apple Harvesting

It can be used to gently pick the fruit from the trees, reducing the risk of damage to the fruit and the need for manual labor.

Fig. 7.19 YOLO V3 and V4 orange detection examples in daytime and nighttime images [5]

[13] This research shows a possible solution for automated apple picking systems: a soft gripper that delivers force feedback and fruit slip detection. The gripper integrates force sensors and slide detection algorithms to provide real-time feedback and improve fruit quality during harvesting, while its soft material and pneumatic actuation ensure a gentle yet firm grasp on the fruit (Fig. 7.21).

[14] This paper provides an adaptive variable parameter impedance control approach for a robotic apple harvesting system, which makes it a promising solution for automated apple picking systems. The proposed control system enables the robot to change its compliance to fit the varied stiffness of the fruit, resulting in delicate but dependable picking. The system's performance was assessed using an apple harvesting robot, and the results revealed that the suggested method surpassed existing methods in terms of harvesting efficiency and fruit quality (Fig. 7.22).

[15] This research presents the potential application of an asymmetric bellow flexible pneumatic actuator in a miniature robotic soft gripper for the purpose of enhancing an automated apple-picking system. The actuator design's compact size and extensive range of motion render it highly suitable for the harvesting of delicate

Fig. 7.20 Examples of apple fruitlet detection results prior to fruit thinning. **a** The detection result of an image acquired in bright and direct sunshine. **b** Detection result of an image captured in both natural and artificial light. **c** Image detection result from cloudy and direct sunshine situations. **d** Image detection result under foggy and backlight situations. *Note* The numbers 1, 3, and 5 correspond to the detection results of apple fruitlets with weak, shadowy, and bright illumination, respectively. The detection result of a single apple fruitlet is 2, and the number of misdetections is 9. The detection results of occluded apple fruitlets, clustered apple fruitlets, apple fruitlets separated into portions by branches or petiole, and fuzzy apple fruitlets are denoted by the numbers 4, 6, 7, and 8, respectively [7]

Fig. 7.21 The soft gripper contact model: **a** initial contact; **b** full contact [13]

Fig. 7.22 The experimental platform designed for the apple harvesting robot involves a grasping mechanism [14]

fruits. The soft gripper was tested on apples, where it showed promise for the future development of soft robotic grippers for automated apple picking systems due to its precision and damage-free object manipulation (Fig. 7.23).

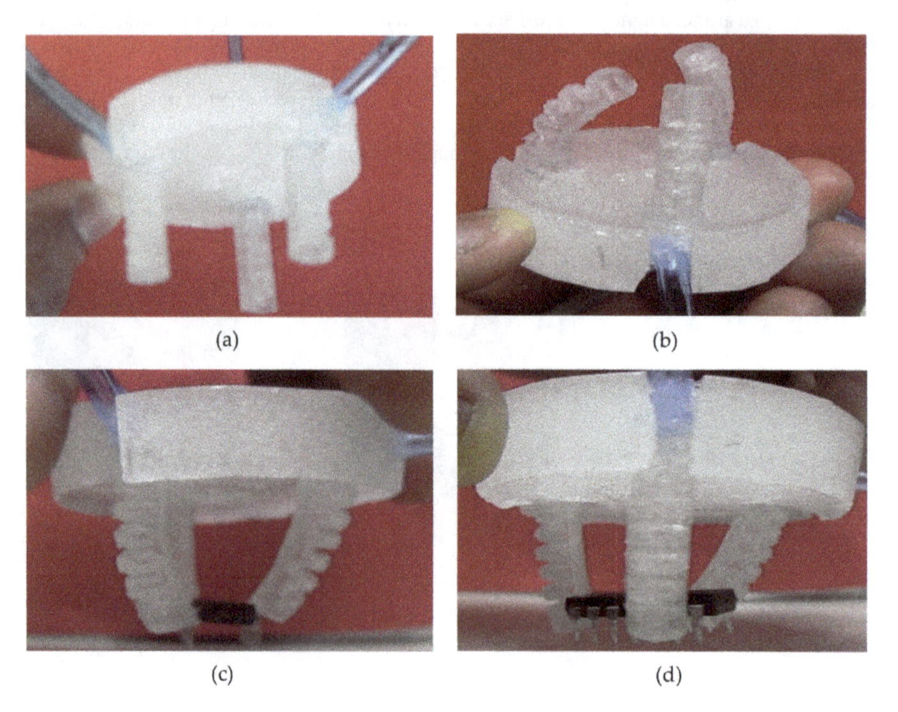

Fig. 7.23 The images depict miniature soft grippers in various scenarios, including: **a** in a state without the application of pressure; **b** in a state with the application of pressure; **c** grasping an IC chip from the front; and **d** grasping an IC chip from the sides [15]

[16] The study on Fruit Classification with Robotic Gripper and Adaptive Grasping can aid the development of Automated Picking Systems for apples. It offers potential to improve accuracy and efficiency in detecting and classifying ripe apples before picking, resulting in reduced damage and increased productivity. The technology could contribute to the development of more precise and reliable harvesting processes in the agricultural industry (Fig. 7.24).

[17] The present research suggests the development of an autonomous fruit harvesting robot that employs deep learning techniques. This innovation exhibits potential for the implementation of robotic arm-based apple picking systems. The automated system employs visual perception through cameras and sensors to detect and recognize fruit, while utilizing a deep learning model to ascertain the most advantageous picking locations and trajectories for the robot arm. The results of the apple tree testing indicated a significant level of success in both the precise identification and selective harvesting of apples (Fig. 7.25).

[18] The present research suggests an economical approach to automate fruit harvesting through the integration of a stereovision camera and a robotic arm. This innovation holds significance for the advancement of apple picking systems utilizing

Fig. 7.24 Illustration of the functionality of the gripper. The manipulator is utilized to seize fruits [16]

Fig. 7.25 Harvest robot [17]

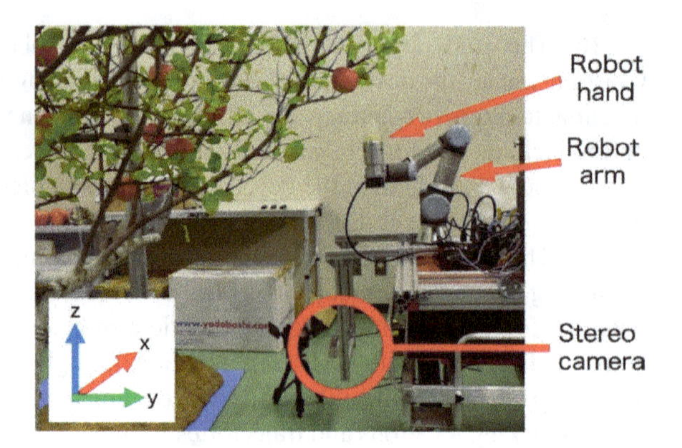

robotic arms. The stereovision camera is employed for the purpose of fruit localization and subsequent calculation of the optimal position for the robotic arm to execute the fruit-picking task. The efficacy of the system was evaluated on apple trees, revealing an 80% success rate in the detection and harvesting of apples. The current research offers a potential resolution for the creation of mechanized robotic arm picking mechanisms for apples by utilizing economical equipment (Fig. 7.26).

[19] The research pertaining to the development of an Apple-Picking End Effector holds significance in the context of the advancement of Automated Picking Robotic Arm Systems for the purpose of apple harvesting. The findings of the study related to the design of end effectors have the potential to enhance the efficacy and precision of robotic arm systems in the selective harvesting of apples. The aforementioned outcome has the potential to facilitate the creation of automated picking systems that

Fig. 7.26 **a** The design of a robotic arm; **b** Elaboration on the gripper tool, which includes an imaging device [18]

are more sophisticated and efficient, not only for apples but also for other types of fruits (Fig. 7.27).

[20] The study concerning the recognition of fruits in real-time and the estimation of grasping is pertinent to the advancement of automated robotic arm systems for apple harvesting. The implementation of a real-time recognition and grasping estimation system in the context of apple picking can enhance the precision and effectiveness of robotic arm systems. The aforementioned could potentially pave the way for the creation of more sophisticated and efficient automated harvesting systems for various types of fruits, including apples (Fig. 7.28).

[21] The research conducted on the Development of a Novel Biomimetic Mechanical Hand, which is based on the physical characteristics of apples, holds significance in the advancement of Automated Picking Suction Cups Systems for apple harvesting. The biomimetic mechanical hand utilized in the study has the potential to enhance the effectiveness and precision of suction cup systems employed in apple

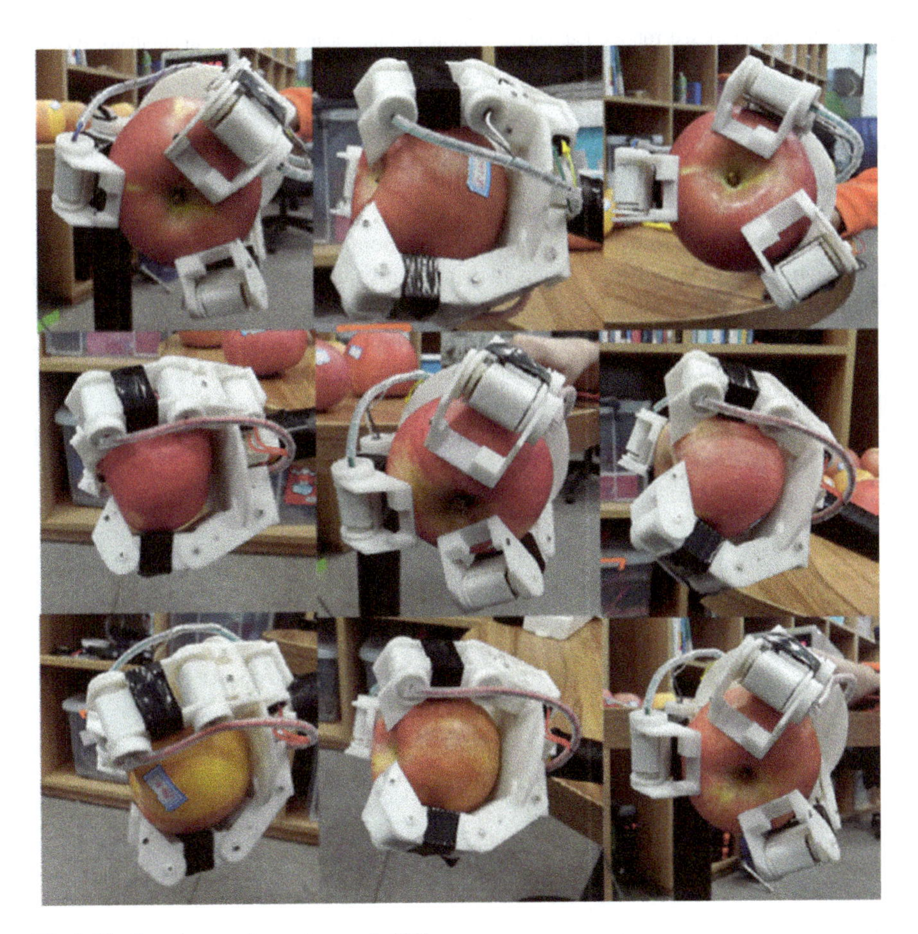

Fig. 7.27 Experiment of grasping apple [19]

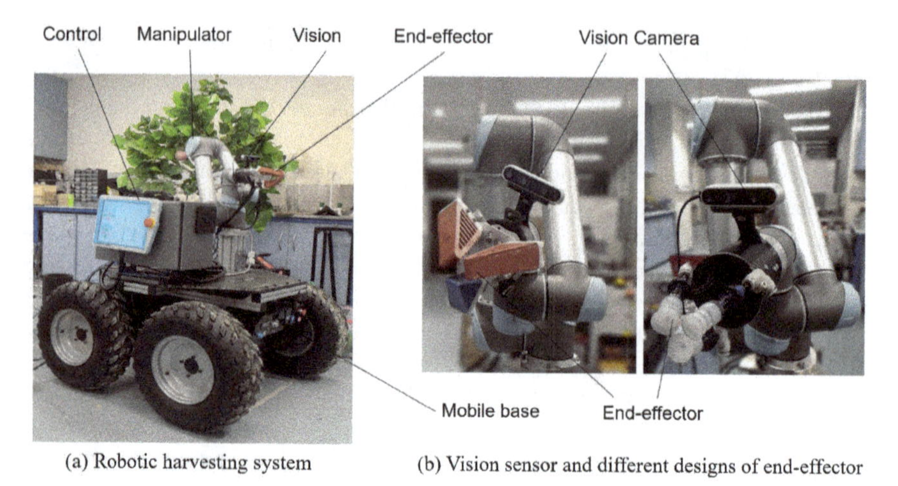

(a) Robotic harvesting system

(b) Vision sensor and different designs of end-effector

Fig. 7.28 The proposed robotic harvesting system includes a mobile base, manipulator, vision camera, and end-effector [20]

harvesting by conforming to the physical attributes of apples. This has the potential to facilitate the advancement and enhancement of automated harvesting systems for fruits such as apples (Fig. 7.29).

[22] The research study involves the development and evaluation of a soft robotic gripper for apple harvesting, which is designed to be robust and gentle enough to avoid damaging the fruit. The gripper was tested on different varieties of apples and shown to be effective in harvesting the fruit without causing any damage (Fig. 7.30).

[23] This research focuses on the different types of end effectors used in agricultural robotic harvesting systems, including grippers, suction cups, and cutting tools. It also discusses the various factors that must be considered when choosing an end

Fig. 7.29 Apple-picking robot equipped with the designed mechanical hand [21]

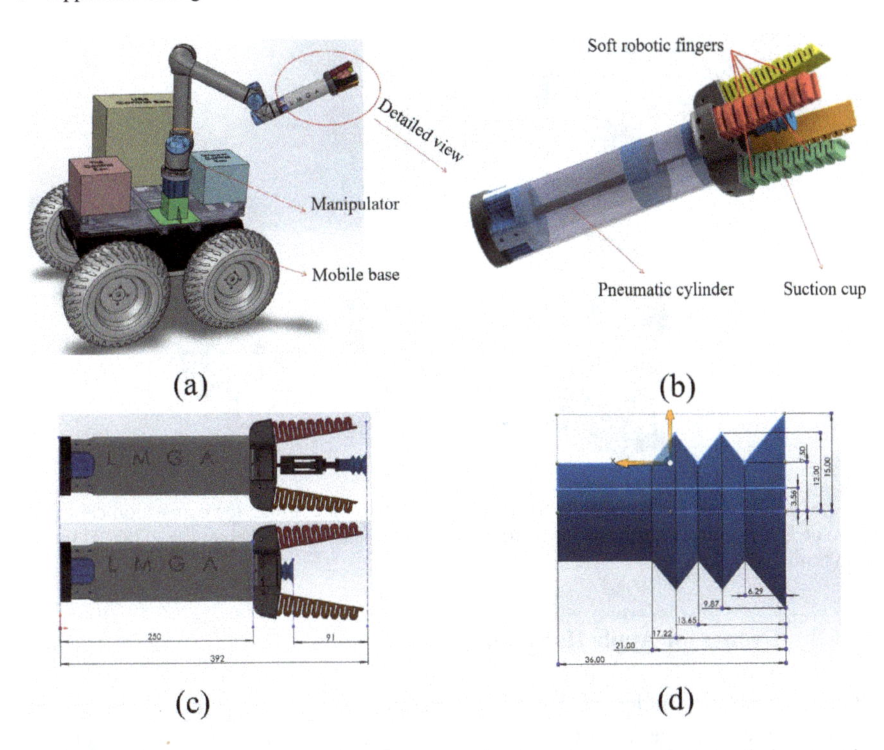

Fig. 7.30 **a** A simplified model has been developed for the entire system. **b** A detailed perspective of the proposed soft robotic gripper is presented. **c** The suction cup's stroke is depicted in this plot, with two SRFs being visualized. **d** The dimensions of the suction cup are provided [22]

effector for a particular crop, such as the size, shape, and delicacy of the fruit or vegetable. It concludes that the choice of the appropriate end effector is critical in agricultural robotic harvesting systems as it significantly affects the system's efficiency and effectiveness. The article proposes an automated system for apple crop harvesting consisting of a robotic complex mounted on a tractor cart. It includes an industrial robot, packaging system, vacuum gripper, vision system, generator, vacuum pump, and equipment control system for high reliability and efficient operation in the field [24]. It is a semi-automatic machine. For semi-automatic technology, typically one worker is needed to drive the machine, and all other work is completed automatically by the machine. [25] A new design of end effectors for grasping round fruits was proposed, adopting a quasi-five-finger structure. In this study, the mechanical strength and finger movement were analyzed and simulated. The control strategy was based on impedance control method, providing good adaptability and effective grasping in complex environments (Fig. 7.31).

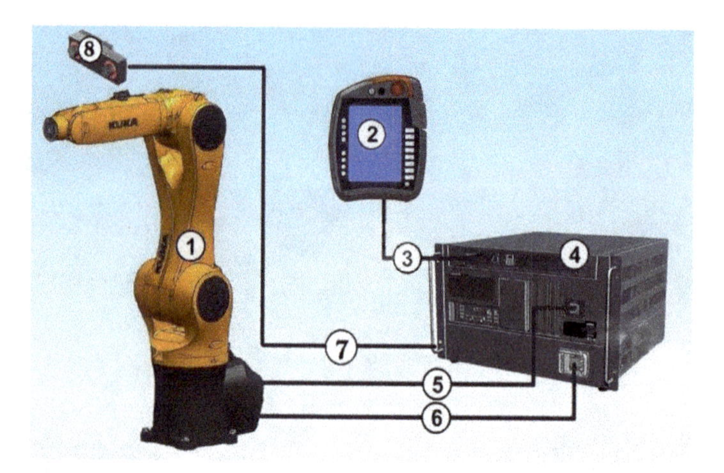

Fig. 7.31 (1) Robotic arm [24]: (2) SmartPAD programming console, (3) connection cable, (4) KUKA KR C4 controller, (5) data cable, (6) motor control cable, (7) Ethernet cable, (8) 3D sensor RC Visard 160

7.1.6.3 Drones for Apple Harvesting

Unmanned aerial vehicles (UAVs) or drones equipped with cameras and sensors can be used to identify ripe fruit on the trees, allowing workers or machines to target only the fruit that is ready to be picked. [26] This study presents the design and development of an unmanned aerial system (UAS) for automated apple harvesting. The UAS consists of a quadrotor drone equipped with a robotic arm and a visual serving system for detecting and picking apples. The system was tested in an apple orchard and demonstrated promising results, with a success rate of 93.4% in picking ripe apples. [27] This study proposes an autonomous unmanned aerial vehicle (UAV) for apple harvesting. The system includes a custom-designed end effector for grasping the apples and a computer vision algorithm for detecting the ripe fruit. The UAV was tested in a small-scale apple orchard and shown to have a success rate of 96% in detecting and harvesting apples. [28] This paper showcases the creation of an independent apple harvesting mechanism that employs a fusion of computer vision, deep learning, and robotic manipulation methodologies, all mounted on an unmanned aerial vehicle (UAV). The apparatus comprises of a quadrotor unmanned aerial vehicle (UAV) that is outfitted with a manipulator arm and a suction cup end effector, which is utilized for the purpose of apple harvesting. The efficacy of the system was evaluated in an apple orchard, and the results indicated a success rate of 86.7% in effectively harvesting mature apples. Overall, modern robotics and machines can improve the speed, efficiency, and safety of apple harvesting. They can also reduce the need for manual labor.

7.1.7 Key Insights from Previous Studies on Apple Harvesting Robotics

Apple Detection and Localization Algorithms: Many techniques, like as YOLOv4, Faster R-CNN, and Mask R-CNN, have been proposed for this task. To identify and locate apples in still images and moving videos, these algorithms employ deep learning and machine learning methods. Some investigations have reported average precision values above 90% in apple recognition and localization, demonstrating their success. Applying these algorithms can enhance apple quality control, inventory management, and sorting.
Methods presented to Boost Accuracy and Efficiency: Several methods have been presented in studies to boost the precision and speed of apple recognition and localization. Feature fusion, channel attention, and spatial attention are only some of the deep learning approaches that can be used to improve object recognition's precision. Color-based segmentation and form analysis are two examples of image processing approaches that can be used to enhance apple detection in challenging settings. To further enhance the effectiveness of identification and localization, machine learning algorithms like decision trees and support vector machines can be applied.
Hardware and Mechanical Solutions: Studies have presented a variety of software solutions for effective apple harvesting and handling in agricultural and industrial contexts. However, studies have also proposed a number of hardware and mechanical solutions. End effectors with a five-finger-like shape have been proposed in research for picking up spherical fruits, and robotic systems mounted on a tractor cart have been proposed for automated apple crop harvesting. These approaches have demonstrated considerable promise in terms of precision, velocity, efficiency, and generalizability to a wide range of settings.

Previous research on robotic apple harvesting has taught us a few important things, primary different algorithms, methods, and hardware solutions can be employed to achieve superior accuracy, efficiency, and adaptability in apple detection and localization. These methods have the potential to boost the efficiency and effectiveness of apple harvesting, sorting, and management in a variety of contexts.

7.1.8 Challenges and Limitations

The development of efficient and effective robotic systems for apple harvesting poses several challenges and limitations that require attention. A significant obstacle encountered during apple harvesting is the intricate surroundings in which it occurs, commonly in orchards characterized by uneven topography, branches, foliage, and various impediments. The navigation of robotic systems in such environments necessitates the avoidance of collisions and damage to both the trees and fruits. Furthermore, apples exhibit significant variations in their shape, size, and color, thereby posing a challenge for robotic systems to precisely detect and grasp them. The adaptability of the systems is crucial for detecting and grasping fruits, even when they are partially obscured by leaves or other objects, given the inherent variability of the fruits.

An additional obstacle pertains to the expenses and capacity for expansion of robotic technologies utilized in the process of collecting apples. The process of creating and implementing such systems can incur significant expenses and require a substantial amount of time, while also potentially lacking the ability to accommodate expansive orchards. It is imperative to conduct a thorough assessment of the cost-effectiveness of these systems to ascertain their economic feasibility. Energy efficiency is a pertinent issue, given that the operation of these systems necessitates a substantial amount of energy. It is imperative that the design of these systems is optimized to maximize energy efficiency while minimizing the duration of downtime required for recharging or battery replacement.

The field of apple harvesting robotics presents supplementary obstacles in the areas of maintenance and reliability. In order to ensure optimal performance and reliability, it is imperative that robotic systems undergo regular maintenance and servicing. The operational systems may necessitate maintenance or substitution of constituent parts, such as grippers, sensors, or motors, thereby augmenting the comprehensive expenditure and intricacy of the systems.

Mitigating these challenges and limitations necessitates additional research and development endeavors in the realm of apple harvesting robotics. Not withstanding, the possible advantages of said systems with regard to enhanced efficiency, productivity, and quality control render them a propitious domain for forthcoming innovation.

7.1.9 Future Directions and Opportunities

The field of robotics in apple harvesting offers numerous prospects and potential avenues for further investigation and advancement. Several prospects are available, which encompass:

The current apple harvesting robotic systems exhibit limitations in accurately detecting and localizing fruits, thereby requiring improvement in this regard. Further studies may concentrate on the enhancement of sensors and machine learning algorithms to achieve greater precision and efficiency in detecting and locating fruits.

The development of novel end-effectors and grippers is a crucial aspect of the advancement of apple harvesting robotics. The current grippers employed in this domain are primarily tailored to facilitate the retrieval of spherical fruits. Prospective studies might focus on the advancement of grippers that exhibit greater adaptability to diverse fruit shapes and sizes, in addition to those that can handle fruits with greater delicacy, thereby mitigating the likelihood of harm.

Enhancing Navigation and Obstacle Avoidance: The ability to navigate and avoid obstacles is of utmost importance in the context of apple harvesting robotics, as it directly impacts the success of the process. Subsequent investigations may concentrate on the enhancement of navigation and obstacle avoidance algorithms that possess the capability to adjust to intricate surroundings and avoid collisions with trees and other objects.

The enhancement of energy efficiency is a crucial factor to be taken into account in the context of apple harvesting robotics, given the substantial energy demand that these systems entail. Prospective research could focus on the advancement of more effective systems that optimize energy consumption and minimize periods of inactivity for the purpose of recharging or replacing batteries.

The present limitation of apple harvesting robotics is their scalability, as they are predominantly designed for smaller orchards. Further research could concentrate on the development of highly scalable systems that are capable of being implemented in expansive orchards and farms, emphasizing the attainment of cost-effectiveness and deployment simplicity.

At present, the majority of apple harvesting robotic systems are only partially automated, necessitating human involvement for specific tasks. The development of fully automated systems is an ongoing area of research and development. Future research might center on the advancement of entirely automated systems that are capable of executing all operations without human involvement, thereby enhancing efficacy and diminishing labor expenditures.

The effective utilization of these prospects necessitates a collaborative effort among researchers, engineers, and industry stakeholders. The apple industry can potentially benefit from enhanced productivity, reduced labor costs, and improved quality control by implementing more sophisticated and efficient robotic systems for apple harvesting.

References

1. Gené-Mola J, Vilaplana V, Rosell-Polo JR, Morros JR, Ruiz-Hidalgo J, Gregorio E (2019) Multi-modal deep learning for Fuji apple detection using RGB-D cameras and their radiometric capabilities. Comput Electron Agric 162(May):689–698. https://doi.org/10.1016/j.compag.2019.05.016
2. Nguyen TT, Vandevoorde K, Wouters N, Kayacan E, De Baerdemaeker JG, Saeys W (2016) Detection of red and bicoloured apples on tree with an RGB-D camera. Biosyst Eng 146:33–44. https://doi.org/10.1016/j.biosystemseng.2016.01.007
3. Gené-Mola J, Vilaplana V, Rosell-Polo JR, Morros JR, Ruiz-Hidalgo J, Gregorio E (2019) KFuji RGB-DS database: Fuji apple multi-modal images for fruit detection with color, depth and range-corrected IR data. Data Br 25:104289. https://doi.org/10.1016/j.dib.2019.104289
4. Li T et al (2021) An improved binocular localization method for apple based on fruit detection using deep learning. Inf Process Agric. https://doi.org/10.1016/j.inpa.2021.12.003

5. Mirhaji H, Soleymani M, Asakereh A, Abdanan Mehdizadeh S (2021) Fruit detection and load estimation of an orange orchard using the YOLO models through simple approaches in different imaging and illumination conditions. Comput Electron Agric 191(June):106533. https://doi.org/10.1016/j.compag.2021.106533
6. Luo L, Tang Y, Lu Q, Chen X, Zhang P, Zou X (2018) A vision methodology for harvesting robot to detect cutting points on peduncles of double overlapping grape clusters in a vineyard. Comput Ind 99:130–139. https://doi.org/10.1016/j.compind.2018.03.017
7. Wang D, He D (2021) Channel pruned YOLO V5s-based deep learning approach for rapid and accurate apple fruitlet detection before fruit thinning. Biosyst Eng 210:271–281. https://doi.org/10.1016/j.biosystemseng.2021.08.015
8. Liu X, Zhao D, Jia W, Ji W, Sun Y (2019) A detection method for apple fruits based on color and shape features. IEEE Access 7:67923–67933. https://doi.org/10.1109/ACCESS.2019.2918313
9. Li T, Feng Q, Qiu Q, Xie F, Zhao C (2022) Occluded apple fruit detection and localization with a frustum-based point-cloud-processing approach for robotic harvesting. Remote Sens 14(3):482. https://doi.org/10.3390/rs14030482
10. Fan P et al (2021) Multi-feature patch-based segmentation technique in the gray-centered rgb color space for improved apple target recognition. Agriculture 11(3). https://doi.org/10.3390/agriculture11030273
11. Ji W, Pan Y, Xu B, Wang J (2022) A real-time apple targets detection method for picking robot based on Shufflenetv2-YOLOX. Agriculture 12(6). https://doi.org/10.3390/agriculture12060856
12. Ji W, Gao X, Xu B, Pan Y, Zhang Z, Zhao D (2021) Apple target recognition method in complex environment based on improved YOLOv4. J Food Process Eng 44(11). https://doi.org/10.1111/jfpe.13866
13. Chen K et al (2022) A soft gripper design for apple harvesting with force feedback and fruit slip detection. Agriculture 12(11). https://doi.org/10.3390/agriculture12111802
14. Wei J, Yi D, Bo X, Guangyu C, Dean Z (2020) Adaptive variable parameter impedance control for apple harvesting robot compliant picking. Complexity 2020. https://doi.org/10.1155/2020/4812657
15. Udupa G, Sreedharan P, Dinesh PS, Kim D (2014) Asymmetric bellow flexible pneumatic actuator for miniature robotic soft gripper. J Robot 2014. https://doi.org/10.1155/2014/902625
16. Zhang J, Lai S, Yu H, Wang E, Wang X, Zhu Z (2021) Fruit classification utilizing a robotic gripper with integrated sensors and adaptive grasping. Math Probl Eng 2021. https://doi.org/10.1155/2021/7157763
17. Onishi Y, Yoshida T, Kurita H, Fukao T, Arihara H, Iwai A (2019) An automated fruit harvesting robot by using deep learning. ROBOMECH J 6(1):2–9. https://doi.org/10.1186/s40648-019-0141-2
18. Font D et al (2014) A proposal for automatic fruit harvesting by combining a low cost stereovision camera and a robotic arm. Sensors (Switzerland) 14(7):11557–11579. https://doi.org/10.3390/s140711557
19. Shi Y et al (2018) Design of an apple-picking end effector. J Mech Eng 64(4):216–224. https://doi.org/10.5545/sv-jme.2017.5084
20. Kang H, Zhou H, Wang X, Chen C (2020) Real-time fruit recognition and grasping estimation for robotic apple harvesting. Sensors (Switzerland) 20(19):1–15. https://doi.org/10.3390/s20195670
21. Wang M et al (2022) Development of a novel biomimetic mechanical hand based on physical characteristics of apples. Agriculture 12(11). https://doi.org/10.3390/agriculture12111871
22. Wang X, Kang H, Zhou H, Au W, Wang MY, Chen C (2023) Development and evaluation of a robust soft robotic gripper for apple harvesting. Comput Electron Agric 204(December 2022):107552. https://doi.org/10.1016/j.compag.2022.107552
23. Vrochidou E, Tsakalidou VN, Kalathas I, Gkrimpizis T, Pachidis T, Kaburlasos VG (2022) An overview of end effectors in agricultural robotic harvesting systems. Agric 12(8). https://doi.org/10.3390/agriculture12081240

24. Krakhmalev O, Gataullin S, Boltachev E, Korchagin S, Blagoveshchensky I, Liang K (2022) Robotic complex for harvesting apple crops. Robotics 11(4):1–15. https://doi.org/10.3390/robotics11040077
25. Shi Y, Yang G, Liu L, Zhao J, Chen J, Cui Y (2016) Design of an end effector for crawling roundlike fruits. Acad J Manuf Eng 14(1):46–54
26. Zhang L, Li Y, Li Y, Li Q, Zhang X, Li H (2020) Design and development of an unmanned aerial system for automated apple harvesting. Comput Electron Agric 175:105524. https://doi.org/10.1016/j.compag.2020.105524
27. Sabzi S, Nourani-Vatani N, Afshar A, Ehsani R (2018) Autonomous UAV for automated apple harvesting. J Intell Robot Syst 91(3–4):583–592. https://doi.org/10.1007/s10846-017-0675-5
28. Lee D, Lee J, Shin D, Choi W, Kim S (2021) Development of a UAV-based autonomous apple harvesting system. Sensors 21(8):2744–2762. https://doi.org/10.3390/s21082744

Chapter 8
Research Advance on Vision System of Apple Picking Robot

Liu Xiaohang, Guo Jiarun, Yang Jie, Afshin Azizi, Zhang Zhao, Dongdong Yuan, and Xufeng Wang

Abstract Vision system is one of the most important components of apple picking robot, which determines the quality and speed of apple picking robot to complete the picking task. To clarify the challenges and the future research directions of the vision system of apple picking robot, this chapter first summarizes the types of vision sensors commonly used in the apple picking robot, and then summarizes the apple target recognition and positioning methods in the vision system of the apple picking robot. The problems existing in the vision system of apple picking robot are further analyzed. It is pointed out that the optimization of the vision system structure, the optimization of the intelligent algorithm in the vision system, the improvement of the real-time performance of the vision system and the improvement of the cost performance of the vision system will become the key research directions in the future, which will provide a reference for further research on the vision system of apple picking robot.

Keywords Visual sensor · Apple detection · Apple positioning

L. Xiaohang · G. Jiarun · Y. Jie · A. Azizi · Z. Zhao (✉)
Key Laboratory of Smart Agriculture System Integration, Ministry of Education, China Agricultural University, Beijing 100083, China
e-mail: zhaozhangcau@cau.edu.cn

Key Laboratory of Agricultural Information Acquisition Technology, Ministry of Agriculture and Rural Affairs, China Agricultural University, Beijing 100083, China

College of Information and Electrical Engineering, China Agricultural University, Beijing 100083, China

D. Yuan
Sweet Fruit, Co., Ltd, Jiangsu Province, Suqian City 223839, China

X. Wang
College of Mechanical and Electrical Engineering, Tarim University, Alar 843300, China
e-mail: wxf@taru.edu.cn

Z. Zhang and X. Wang (eds.), *Towards Unmanned Apple Orchard Production Cycle*, Smart Agriculture 6, https://doi.org/10.1007/978-981-99-6124-5_8

8.1 Introduction

China is the world's largest producer and consumer of apples, with its planting area and output accounting for more than 50% of the world's total, making it a key player in the global apple industry and contributing significantly to its gross national product [1]. However, due to the complex and variable ecological environment of the orchard, it still relies on manual picking at present, which has significant drawbacks, including low efficiency and high cost. Due to factors such as labor shortages, rising labor costs, and an aging workforce, the proportion of apple harvesting cost in the total production cost will rise further, which poses a serious threat to the economic benefits of fruit farmers and also brings great obstacles to the healthy and sustainable development of the Chinese apple industry [2]. Hence, the development of an apple-picking robot to replace manual picking is crucial, as it holds immense practical value and economic benefits. Apple picking robots are primarily composed of robotic arm system and visual system. The robotic arm performs apple picking under the guidance of the visual system. Therefore, rapid and accurate identification and positioning of apples is the key to realize automatic picking [3]. In this chapter, on the basis of discussing the hardware of the vision system of apple detection, two methods of apple detection are summarized and analyzed. At the end, this chapter identifies the key research directions for detecting apples in the future.

8.2 Sensors for Fruit Detection

The visual system serves as the foundation for the apple-picking robot to locate and collect apples. However, its accuracy in fruit detection is affected by unpredictable and variable lighting conditions in the field environment, varying and complex canopy structure, and diverse colors, shapes and sizes of fruits [4]. In addition, the accuracy of fruit detection is also limited by the fact that the fruit in the crown image is blocked by leaves, branches and other fruits [5]. Therefore, researchers attempt to enhance the accuracy of fruit detection by utilizing cameras that can capture various image information. The various types of cameras used for apple's testing are shown in Table 8.1.

8.2.1 Black and White Camera

Black and white cameras were used in early studies of fruit detection based on geometric features. Researchers demonstrated the feasibility of a black and white camera in a self-propelled prototype robot they developed for apple harvest [6]. Later, Cardenas-Weber [7], Dobrusin [8], Pla [9], Edan [10], and others employed black and white cameras to complete the detection of melon, citurs, and other fruits

Table 8.1 Camera types used for apple target detection

Sensors	Advantages	Limitations
Black and White camera	Minimal effect of variable lighting conditions	Lack of color information
Color camera	Provides color, geometric and texture information	Affected by lighting conditions
Spectral camera	Can have both spectral and color information	Time consuming
Thermal camera	Independent of fruit color	Affected by size of fruit; narrow range of operations during day time
Depth camera	Provides three-dimensional information and accurate positioning	Cost high

based on the geometric shape and texture of fruits, but the resulting accuracy was not high. To this end, the researchers attempted to introduce color filters into the black and white camera to provide more useful information for fruit detection, and achieved an 80.0% accuracy and 10.0% false detection rate. This attempt confirmed the importance of color information, and subsequently black and white cameras were replaced by color cameras and achieved a 90.0% accuracy with a 5% error detection rate. Black and white cameras are limited because they lack access to the critical characteristic of fruit-color information- even though they can effectively reduce the impact of complex and variable lighting factors.

8.2.2 Color Camera

Color cameras with CCD or CMOS sensors have become the most widely used cameras in robotic vision and automated agriculture because of their low cost. In addition to providing geometric and textural information, color cameras also provide opportunities for fruit segmentation based on color information. For instance, D. M. Boanon [11, 12] and Si Yongsheng et al. [13] effectively distinguished background apples from target apples based on the color threshold segmentation method. However, their detection results were affected by the color change of fruits during ripening. Hence, they could not detect fruits of different maturity and specific varieties at the same time. In addition, cloudy or sunny weather and direct sunlight also pose great challenges to fruit detection methods based on color characteristics. Therefore, color cameras need to overcome their sensitivity to illumination changes before making full use of color information.

8.2.3 Spectral Camera

Spectral imaging technology is used to detect objects by the absorption or radiation properties of substances on different electromagnetic spectrum. A spectral camera adds one-dimensional spectral information along the spatial direction to the common two-dimensional spatial imaging, which makes it useful in various research fields, including apple detection. Compared with color cameras, spectral cameras have an advantage in apple detection because they can separate fruit and background even when they are similar by analyzing their spectral information, providing more accurate damage judgments. For example, the researchers used the obtained multi-spectral and hyperspectral image data combined with principal component analysis dimension reduction, morphological transformation, watershed, and spot segmentation algorithms to achieve apple segmentation with an accuracy of 88.1%. However, the target movement caused by stroke and the change of shadow position between different filter images have great influence on the final results. In addition, the feasibility of spectroscopic cameras for fruit detection is limited by the time-consuming image acquisition and analysis process.

8.2.4 Thermal Camera

Thermal cameras can capture the temperature signature of an object, which helps distinguish fruit from background. Fruits absorb and radiate more heat than leaves and other parts of the plant canopy, allowing thermal imaging to distinguish those plant materials. Researchers have used thermal imaging to detect green fruits that are difficult to identify using color information alone, based on the temperature difference between the fruit and the background. For example, Stajnko et al. [14] used thermal images to detect apples by taking advantage of the differences in the rates of heat absorption and radiation of fruits, leaves and branches. In order to maximize the temperature difference between target apples and the background, thermal images were collected in the afternoon, and the correlation coefficient between the number of apples measured by eye and the number detected by thermal images was finally 0.85. Bulanon et al. [15] demonstrated that combining thermal and RGB images outperformed using only a single thermal image for fruit detection. The fusion approach achieved a higher detection accuracy than the single thermal image approach. However, the accuracy of the thermal imaging method is limited by fruit size and direct sunlight. Accuracy is lower when detecting fruit that is shaded and deep into the canopy, as there is no significant temperature difference between the fruit and the background in these areas. In addition, since the time of image acquisition during the test in a day will directly affect the image quality, the author suggests adding shape features to the fruit detection method to improve the accuracy of the system by identifying partially hidden fruits.

8.2.5 *Depth Camera*

Depth cameras are capable of capturing both color and depth images. Color image can provide two-dimensional information such as color, texture and shape of the target, while depth image contains three-dimensional spatial information of the target. The fusion of the two can effectively remove the invalid background and effectively improve the detection accuracy of the target fruit. For example, Fu et al. [16] used the depth image obtained by the Kinect V2 depth camera along with artificial threshold to eliminate the irrelevant background (non-target tree and fruit) in the color image, and Faster R-CNN network was used to detect fresh fruit. The results showed that removing the background with the depth threshold improved the detection accuracy of apple by 2.5%. Furthermore, there has been a significant improvement in the average precision value comparing with the original RGB image. Mai Chunyan et al. [17] performed three-dimensional reconstruction of fruit trees by integrating the color and depth information provided by the RGB-D camera. They employed R-G color difference threshold segmentation and filtering denoising techniques on the fruit tree point cloud to obtain the point cloud information of the fruit region. Finally, the point cloud segmentation method with random sampling consistency was used to extract the three-dimensional sphere shape of the fruit point cloud to obtain the three-dimensional spatial position information of each fruit. Although these studies have proved the effectiveness of depth camera, most of them remain in the preliminary utilization of 3D information, without deep fusion of real and RGB image information, which is also the key reason why apple detection algorithm is so difficult to achieve important breakthrough.

8.3 Traditional Apple Detection Methods

At present, many machine learning and image processing methods are widely used to detect apples in apple orchards. These methods usually use color threshold segmentation, morphological transformations, edge detection, circular Hough transform, region growth, and other operations to preprocess the image, and then manually design a feature extractor to obtain the color, texture, shape, size, and other different apple features and their combinations. Finally, the collected features are fed into machine learning models, such as support vector machines (SVMs), k-nearest neighbors (KNNs), and k-means clustering algorithms. This allows the trained model to detect apples in unseen images [18]. However, most of these methods are targeted at scenes and use light-sensitive RGB images, which have poor robustness and generalization in the changeable natural environment. Balancing real-time and accuracy can be challenging, often leading to missed or false detections. Therefore, it cannot meet the actual work requirements of the picking robot in a variety of complex scenes.

8.3.1 Color Features

In an orchard environment, fruit color is one of the most important features for machine vision systems to distinguish apples from leaves and branches. Some researchers have been able to detect apples with 92% accuracy based on color features alone. Bulanon et al. [12] proposed a method to detect red Fuji apples under different lighting conditions by using the red threshold value. Their results showed that the detection accuracy of apples was strongly correlated with the lighting conditions, and the detection accuracy would be greatly improved under controllable lighting conditions. Zhou et al. [19] used RGB and HSI as color spaces for apple detection task (Fig. 8.1). First, red and green apples separated from the background with the help of RGB three-channel images, and then the saturation channel was used to segment red apples. The regression coefficients of red and green apple counts were 0.80 and 0.85, respectively. The accuracy of this method is strongly affected by the shading of branches, leaves, and trunks, as well as the clustering of fruits. Zhang Chunlong et al. [20] collected green apple tree images under ring flash light at night. They used the normalized G component, H, and S components in HSV space and super green operator (2G-R-B) as features to recognize fruit tree images reaching the recognition accuracy was 89%. In addition, the red and green difference feature of apple image is also used for apple detection. However, most of the above detection methods based on color characteristics are only applicable to the detection of red ripe apples, and the detection performance with different maturity or specific varieties (Australian Qingping, Yellow Marshal) is poor, and the complete apple region cannot be accurately detected. In addition, both sunny and cloudy weather conditions, as well as whether the fruit is exposed to direct sunlight, can pose significant challenges for fruit detection methods that rely on color features. This is because color cameras, which provide these features, are highly sensitive to changes in lighting.

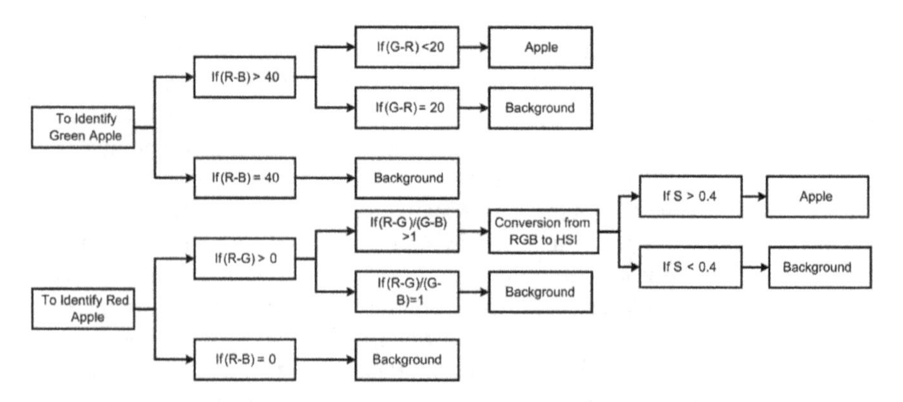

Fig. 8.1 Flowchart explaining different steps of fruit segmentation (Adopted from Zhou et al., 2012) [19]

8.3.2 Geometric Features

Geometric features, such as shape and size, provide an alternative solution for fruit recognition. Usually, the shape of the fruit is similar, and their shape is quite different from that of the branches and leaves. When the color of the fruit is not very different from the background, the geometric feature of the fruit shape is often more effective. Compared with color features, such geometric features tend to be less sensitive to light and are therefore more suitable for real orchard environments. For example, based on the probabilistic Hough transform theory, some researchers proposed a general method to detect fruits by using partial geometric contour feature information and complete fruit classification by SVM algorithm. Feng Juan et al. [21] simplified the three-dimensional images of fruit trees obtained by laser a vision system into binarized images. They achieved a high recognition rate of apples on trees in natural scenes by using target contour extraction and the random ring method. Tsoulias et al. [22] adopted LiDAR to propose an apple recognition method based on the reflectivity and geometric features of Lidar, which could alleviate the difficulty of apple recognition under different illumination, but the processing time of this method was long. Sun Sashuang et al. [23] segmented a single apple object blocked by branches in natural scenes, and obtained the reconstructed image of the blocked apple by fitting the contour curvature features with the three-point fixed circle method, which greatly improved the recognition rate of the blocked fruit. The recognition effect of the target blocked by apple was not ideal, and the average reconstruction time was long. However, the main problem of this method is that when the fruit is blocked by leaves, branches and other fruits, the changes of fruit shape, size make it become difficult to obtain better adaptability in batch processing, and the detection results mostly depend on the results of previous pretreatment outcomes. In addition, most of these algorithms are computationally intensive, which limits their application in real-time environments.

8.3.3 Texture Features

Grain is another distinguishing feature that separates the fruit from the background. The surface of the fruit is usually more fine and smoother than that of the leaves and stems. Fruit recognition based on texture features can still be applied when fruit color is similar to that of leaves and branches, because fruit color does not affect the extraction of texture features. For example, Feng et al. [24] proposed an apple recognition algorithm for multi-spectral dynamic image analysis. This method uses a forward-looking infrared camera with multi-spectral dynamic imaging technology to collect images of fruit. The imaging principle is used to obtain the maximum contrast between the fruit and the background. Then, the method uses the texture information of the fruit to achieve regional localization. The results show that the algorithm is good for detecting unshielded apples, but due to the influence of temperature difference between fruits and branches and leaves, the detection accuracy of shielded apples needs to be improved. Furthermore, the study identified variable lighting conditions as a limitation for texture feature detection in orchards, since it can affect the fruit's texture properties.

8.3.4 Integration of Color, Geometric and Texture Features

The performance of fruit detection methods based on color and texture features is affected by varied light conditions in orchards, while the extraction of fruit geometric features is hindered by fruit overlap and occlusion. Therefore, fruit detection methods based on a single feature (color, texture or geometry) are difficult to achieve the best results. Therefore, many researchers try to combine various characteristics to improve the accuracy and robustness of fruit detection methods. Zhao et al. [25] used orchard images with both red and green apples as their research object. They combined red and texture-based edge detection, and used the Laplacian operator as a filter to separate fruits and leaves. They achieved a recognition success rate of 90%. Wu et al. [26] used SVM to integrate HSV fruit color and 3D geometric features to recognize fruits, and experiments showed that the fruit recognition accuracy of this method was 80.1%. Liu et al. [27] proposed a detection method for apples that utilizes the color and shape characteristics of apples. The study achieved a recall rate of 89.80% and a precision value of 95.12%. However, this algorithm is easily affected by noise and it takes 1.94s to detect an image. Linker et al. [28] detect pixels that may belong to apples through color and texture, form seed regions by these pixels, and then compare them with ideal feature models to determine whether the region contains apples. Liu Lijuan et al. [29] conducted threshold segmentation of the image through Lab and YUV color spaces, and then used circular constraint method and Hough algorithm to detect the fruit, thus realizing the location and segmentation of apple. Jana et al. [30] proposed a multi-type fruit classification algorithm based on texture and color features. This method uses GrabCut image segmentation method to segment fruit targets and background regions in the image. Then GLCM is used to calculate the scalar features of texture attributes (contrast, correlation, energy and uniformity) along horizontal, vertical and diagonal directions. Then the four texture and color features are combined as the feature input of SVM classifier. The accuracy of the trained classifier for apple, pear and cucumber were 91.67%, 75.00% and 83.33%, respectively. Xu et al. [31] proposed an apple recognition and localization method combining PCNN image segmentation and SUSAN edge detection. To reduce the effect of natural light on apple image segmentation, a homomorphic filtering image enhancement algorithm is first used to enhance the image. Then, the R-G factor is used to process the image in RGB color space. Finally, the improved random Hough transform method is used to obtain apple feature circles in the image according to the shape characteristics of the apple. Finally, SUSAN edge detection algorithm was used to obtain the contour information of apple, so that the detection accuracy of 50 apple images was 93%. Gongal et al. [32] carried out histogram equalization processing on color images in

HSI color space, and applied pixel-level adaptive Wiener filter and Otsu threshold algorithm to remove noise and perform rough segmentation of both background and apple. Then, the apple edges found by Canny edge detector were combined with circular Hough transform and blob analysis to realize apple recognition. The method achieved 82% accuracy in estimating fruit yield with bilateral images. Si Yongsheng et al. [33] selected R-B as color features, gray mean, standard deviation, and entropy as texture features, and used K-means clustering method to cluster and segment green apple images, and their apple recognition accuracy reached 81%. Kurtulmus et al. [34] trained and recognized green apple tree images under natural light by using "Eigenfruit", color and Gabor texture features, and the recognition accuracy was 75.3%. Rakun et al. [35] used spatial frequency, texture and multi-view geometric features to segment and recognize 4 groups of 3D green apple images in natural scenes under backlight, and calculated that the proportion of apple pixels recognized were 3%, 1%, 66% and 33%, respectively. Zania et al. [36] proposed a fruit detection method based on key point candidate and random forest classifier. Aiming at problems such as similar color and partial occlusion between fruits and branches and leaves, combined with improved directional gradient histogram and texture intensity descriptor features, random forest algorithm was adopted for classification, and a corner invariant maximum detector was proposed for fruit detection. It has certain scale invariance in both apple and grape datasets, but the detection accuracy needs to be improved. Liao Wei et al. [37] proposed a random forest green apple recognition method based on Otsu threshold segmentation. Due to the similar color of apple fruits and leaves, complex background color and texture features of branches, accurate recognition can be affected. Otsu threshold segmentation and filtering processing are utilized on apple tree images in the RGB color space. The image containing only fruit and leaves is segmented, and the gray scale and texture features of the segmentation image are extracted. The recognition model is constructed by using random forest algorithm, which can effectively use the segmentation map for recognition. The model has certain robustness, and the recognition accuracy reaches 88%. Tao et al. [38] proposed an improved feature extraction method that extracts RGB and HSI color components from point cloud data. They then combined these components with the three-dimensional descriptor FPFH to describe the three-dimensional geometric features of apples. This improved the robustness of the method based on color features. The support vector machine optimized based on genetic algorithm was used to automatically identify apples, branches, and leaves in the scene. Jia et al. [39] used the pulse-coupled neural network to extract 16 typical features of target apples, including 6 color features and 10 shape features, and established a new GA-Elman algorithm to recognize overlapping apples, achieving a recognition rate of reaching 88.67% respectively, which improved the operation efficiency and recognition accuracy. The method described above can successfully identify fruits from simple images. However, in the real application scenario, as shown in Fig. 8.2, the complex background, changing lighting conditions, and common phenomena such as overlap and occlusion bring great challenges to the traditional method. Hence, the performance of the traditional method will also decline sharply.

Fig. 8.2 Apple images in different scenes

8.4 Apple Detection Methods Based on Deep Learning

In recent years, with the rapid development of artificial intelligence, compared with traditional methods, the target detection algorithm based on deep convolution neural network shows great advantages. Deep convolution neural network can directly learn the high-dimensional abstract features of a specific target from a large number of samples, which is more robust than the features extracted manually [40]. At present, apple target detection algorithms based on deep learning are mainly divided into two categories. One is a two-stage detection algorithm represented by RCNN series algorithms. The core idea is to generate a series of candidate frames, and classify and regress the targets after screening the candidate frames. The other is a single-stage detection algorithm represented by SSD and YOLO, which directly regresses the position of the target frame without regional suggestion.

8.4.1 Two-Stage Apple Detection Methods

In actual production, the complex environment of orchard greatly affects the recognition rate of apples. In order to solve this problem, researchers put forward a method of using two-stage target detection network with more accuracy advantage for apple recognition. Li Linsheng et al. [41] took the apple image under natural light as their research object, and put forward an apple target detection model based on Faster RCNN, achieving the accuracy of apple detection 97.6%. Chu et al. [42] improved the detection accuracy of apples and achieved faster detection speed by adding suppression branches to the Mask RCNN. This filtered out the non-apples features learned by the backbone network and resulted in an F1 value of 0.905. Jing Weibin et al. [43] proposed a Faster R-CNN recognition algorithm based on VGG16, and the recognition accuracy was 91%. Gao et al. [44] used the improved Faster R CNN network to detect apples in dense-leaved fruit trees. The mean average precision (mAP) was 87.9%, and the average detection time of a single image was 0.241 s. Jia et al. [45] combined the residual network (ResNet) with the dense convolution network (DenseNet) on the basis of the Mask RCNN network, and realized the accurate detection of apples. Wang et al. [46] developed an accurate apple instance segmentation method based on improved Mask RCNN, which fused an attention module into the backbone network to improve its feature extraction ability. Jia et al. [47] proposed a segmentation algorithm for overlapping green apples called RS-Net. This algorithm extends Mask R-CNN and incorporates an attention mechanism to focus on informative pixels while suppressing noise from occlusion and overlapping, making it more suitable for operation in complex natural environments and increasing its robustness.

8.4.2 One-Stage Apple Detection Methods

In order to meet the real-time requirements of apple target detection, a one-stage detection method with more speed advantage is used for apple recognition. Chen et al. [48] developed an apple detection method based on Des-YOLOv4, which used the Soft-NMS instead of NMS to solve the problem of missed detection, resulting in improved accuracy for fruit detection. Mao Tengyue et al. [49] proposed an improved YOLOv3 apple detection algorithm. The algorithm introduced multi-scale convolution and an attention mechanism, which strengthened the extraction of important network feature information. This resulted in more accurate detection of apple targets. Ma et al. [50] proposed an improved RetinaNet, using MobileNetv3 as the feature extraction network, improved the feature pyramid network, and optimized the size of the anchor frame by using K-means clustering algorithm, which improved the accuracy and speed of apple detection by the network. Yue Linqian et al. [51] proposed an apple detection model based on improved YOLOv4. This model introduces attention mechanism and path aggregation network to enable fruit detection and estimation of fruit diameter. In the methods mentioned above, researchers introduced visual attention mechanisms such as convolutional block attention module (CBAM), Squeeze-and-Excitation block, and Non-Local block into the model. These techniques have improved detection accuracy, but they have also increased the size of the models. Therefore, Zhuo Wang et al. [3] proposed a lightweight real-time detection method for apples (YOLO v4-CA) based on YOLOV4. This method uses MobileNetv3 as the feature extraction network, and introduces deep separable convolution into the feature fusion network to reduce the computational complexity of the network. At the same time, in order to make up for the precision loss caused by model simplification, coordinate attention mechanism is introduced in key positions of the network to strengthen target attention to improve the ability of dense target detection and anti-background interference. The experimental results show that the average detection accuracy of the improved model is 92. 23%, and the detection speed on the embedded platform is 15.11 f/s, which is faster that of the original model. Wu Xing et al. [52] improved the original YOLOv3 by building a backbone network with isomorphic residuals in series, optimizing the loss function, and trained the model through multi-stage learning optimization. The experiment yielded a detection accuracy of 94.57% for apples on fruit trees. However, the study focused on a relatively narrow range of fruit statuses and did not account for objective factors such as changes in illumination, bagging, or low light conditions at night, which ultimately limited the scope of the findings. Gao Fang et al. [53] proposed a fast detection and tracking method for apples based on YOLOv4-tiny lightweight network and Coleman filter algorithm, and realized fruit counting by improving Hungarian Kalman filter. Li Nannan [40] proposed an apple detection model based on Slim-FCOS lightweight network. The model takes DarkNet19 as the backbone network and uses path aggregation network instead of feature pyramid network. At the same time, in order to further simplify the network model and ensure the detection efficiency, FCOS is pruned by using BN-based channel pruning algorithm to reduce the computational complexity. Finally, the trimmed model is fine-tuned by using CIoU border regression loss, and the rapid and accurate detection of apple fruits is realized. The Slim-FCOS detection model can effectively detect fruits under varying illumination and density, with an average accuracy of 95.4%, as demonstrated by the experimental results. Compared with FCOS network, the model size and parameters decreased by 70.2 MB and 62.3% respectively.

In addition, in order to meet the needs of fruit yield measurement and maturity detection at different growth stages in actual production, Lu et al. [54] proposed an improved YOLOv4 apple detection model, which improved the detection accuracy by adding convolution attention module, and realized the detection of unripe and ripe apples. Zhao Dean et al. [55] achieved the detection of bagged, immature and mature apples in complex environment based on YOLOV3 model, with an average detection accuracy of 87.71% and a detection time of 16.69 ms. Tian et al. [56] proposed an improved YOLOv3 model to detect apples at different growth stages in orchards, and enhanced the use of features by adding dense connection structures in the low-resolution feature layer of YOLOv3 network. The average F_1-score of the improved YOLOv3 model for detecting young apples, growing apples and mature apples was 0.817, and the average time for detecting an image was 0.304 s, which had a certain guiding role for apple yield estimation in the early orchard. Zhao Hui et al. [57] (2021) proposed an improved fruit identification method using YOLOv3, which enables the identification of different ripe apple fruits in complex orchard environments with an improved detection accuracy of 96.3%. Fruit recognition is the first step in apple picking robot. However, due to the limitations of the robot's end effector, which can only pick fruits that are not covered or are minimally covered by leaves, it is necessary to detect the fruit state in multiple categories to avoid damage to the mechanical picking arm and end effector. LV et al. [58] proposed an apple growth morphology recognition method based on YOLOv5. While this algorithm boasts an average accuracy of 98.45% and a processing speed of 71 frames/s, it should be noted that it classifies the growth morphology of multiple apples as a whole rather than recognizing the growth state of individual apples. YAN et al. [59] put forward an apple recognition model based on the improved YOLOv5s, which can be divided into two categories: pickable and non-pickable. The model has a recognition accuracy of 83.83% for occluded apples, with an average recognition time of 15ms per image. Therefore, it may miss some apples. In order to solve the above problems, Yan Bin et al. [60] put forward an apple recognition algorithm based on improved

YOLOv5m under different conditions of branch occlusion, so that the manipulator can change the picking posture and make it possible to pick apples under branch occlusion. This algorithm has an 81% recognition accuracy with an average recognition time of 25 ms per image. In order to improve the recognition accuracy, GAO et al. [44] proposed a multi-class apple detection method for dense-leaved fruit trees based on Faster R-CNN. The average accuracy is 87.9%, and the average speed of processing images is 241 ms. Sun Jun et al. [61] proposed an improved RetinaNet-based method for detecting apple occlusion in complex orchard environment. The accuracy of apple detection was analyzed with respect to leaf occlusion, branch/wire occlusion, fruit occlusion, and no occlusion. The results showed that the detection accuracy was 94.02%, 86.74%, 89.42%, and 94.84% respectively. The average accuracy across all scenarios was found to be 91.26%, with a single frame processing time of 42.72 ms. Due to imbalanced bid numbers across different types of occlusion, classification deviations or algorithmic performance degradation may occur. Li et al. [62] constructed eight types of occlusion Apple (MTOA) datasets, and then proposed a balance enhancement method to improve the average detection accuracy of the lightweight target detection model.

Based on the above analysis, it is evident that Faster R-CNN, as a two-stage target recognition algorithm, delivers high recognition accuracy. However, its processing speed is relatively slow, which falls short of meeting real-time requirements. Among them, MaskR-CNN combines target detection with instance segmentation to improve the recognition and positioning accuracy of occluded and overlapping apples. The YOLO algorithm, a single-stage target recognition system, strikes a balance between accuracy and speed. However, it may not perform well in recognizing small target objects due to lower accuracy. The anchor-frame-free recognition algorithms such as CenterNet have high recognition accuracy and speed for dense and occluded apples. Although deep learning algorithms have made significant progress in apple recognition, certain challenges remain, including extended training times, low accuracy of small sample apple recognition, and high computing resource demands. One particularly challenging aspect is identifying occluded apples, which has yet to yield satisfactory results and requires further research and development. In the future research of apple identification methods, combining identification algorithm with identification strategy would be an effective way to address this issue.

8.5 Summary and Prospect

Numerous scholars, both domestically and internationally, have conducted extensive research on apple detection and showcased its potential applications. Despite these advancements, there are still some persisting issues related to the accuracy and real-time performance of the identification algorithm. Due to the complexity of the apple growing environment, the existing algorithms cannot achieve a good balance among recognition accuracy, adaptive selection of parameters and running time. The practicability, real-time performance and stability of existing algorithms in dealing with uncertain illumination, apples of the same color, occlusion, overlap, shadows and other external factors need to be improved. Although there are several apple recognition algorithms available, they tend to focus on specific situations, whereas the natural environment is constantly changing. Therefore, there is a need to improve the universal applicability of these algorithms.

Given the aforementioned problems, we propose the following solutions: (1) Because illumination is the main factor affecting the accuracy of the model, it is expected to reduce its influence by adding active light sources and shading devices to the vision system. (2) To enhance the accuracy and efficiency of apple detection in natural environments, the deep learning algorithm requires further optimization. This can be achieved through several methods such as improving the dataset to increase the accuracy or speed of apple recognition, enhancing the feature extraction module of the algorithm, adding the visual attention mechanism module, and simplifying the model. (3) Creating an open dataset that includes multiple apple varieties, trees, and environmental scenes would greatly enhance the accuracy and speed of apple identification in complex environments. Such a dataset would be a valuable tool for improving the performance of apple detection algorithms.

References

1. Meng Xiangning, Zhang Zihan, Li Yang, Ren Longlong, Song Yuepeng (2019) Research status and progress of apple classification [J]. Deciduous Fruit Tree 51(6):24–27
2. Bai Jinhua (2022) Apple detection in orchard based on computer vision in natural environment [D]. Guizhou Minzu University
3. Zhuo Wang, Wang Jian, Wang Xiaoxiong et al (2022) Lightweight detection method of apples in natural environment based on improved YOLO v4 [J]. J Agric Mach 53(8):294–302
4. Manoj K, Qin Z et al (2012) Mechanization and automation technologies in specialty crop production [J]. Resource Eng Technol Sustainable World 19(5):16–17
5. Gongal A, Silwal A, Amatya S et al (2015) Apple crop-load estimation with over-the-row machine vision system. Comput Electron Agric 120:26–35
6. D'Esnon AG, Rabatel G, Pellenc R, Journeau A, Aldon MJ (1987) Magali: a self propelled robot to pick apples. ASAE Paper 87–1037, St. Joseph, MI
7. Cardenas-Weber M, Hetzroni A, Miles GE (1991) Machine vision to locate melons and guide robotic harvesting. ASAE Paper Number: 91–7006, St Joseph, MI
8. Dobrusin Y, Edan Y, Grinshpun J, Peiper UM, Hetzroni A (1992) Real-time image processing for robotic melon harvesting. ASAE Paper No. 92–3515, St. Joseph, MI
9. Pla F, Juste F, Ferri F (1993) Feature extraction of spherical objects in image analysis: an application to robotic citrus harvesting. Comput Electron Agric 8(1):57–72
10. Edan Y, Rogozin D, Flash T, Miles GE (2000) Robotic melon harvesting. IEEE Trans Rob Autom 16(6):831–835
11. Bulanon DM, Kataoka T, Okamoto H et al (2004) (2004) Development of a real-time machine vision system for the apple harvesting robot [C]// Sice. Conference 1:595–598
12. Bulanon DM, Kataoka T, Ota Y et al (2002) AE—automation and emerging technologies: a segmentation algorithm for the automatic recognition of Fuji apples at harvest. Biosyst Eng 83(4):405–412
13. Si Yongsheng, Liu Gang, Gao Rui (2009) Segmentation algorithm for green apples recognition based on K-means algorithm [J]. Transactions of the CSAM 40(S):100–104. (in Chinese with English abstract)

14. Stajnko D, Lakota M, Hocˇevar M (2004) Estimation of number and diameter of apple fruits in an orchard during the growing season by thermal imaging. Comput Electron Agric 42(1):31–42
15. Bulanon DM, Burks TF, Alchanatis V (2008) Study on temporal variation in citrus canopy using thermal imaging for citrus fruit detection. Biosyst Eng 101(2):161–171
16. Fu L, Majeed Y, Zhang X et al (2020) Faster R-CNN–based apple detection in dense-foliage fruiting-wall trees using RGB and depth features for robotic harvesting [J]. Biosys Eng 197:245–256
17. Chunyan M, Lihua Z, Hong S et al (2015) Three-dimensional reconstruction of fruit trees and fruit identification and location based on RGB-D camera [J]. J Agric Mach 46(S1):35–40
18. Liu Yizhen (2022) Research on apple detection and segmentation based on deep learning [D]. Shandong University.
19. Rong Zhou, Lutz Damerow, Yurui Sun et al (2012) Using colour features of cv. Gala' apple fruits in an orchard in image processing to predict yield. Precision Agric 13(5):568–580
20. Zhang Chunlong, Zhang Ji, Zhang Junxiong et al (2014) Identification method of green apples on trees in near-color background. J Agric Mach 45(10):277–281
21. Feng Juan, Liu Gang, Si Yongsheng et al (2013) Apple recognition algorithm on trees based on laser scanning 3D images [J]. J Agric Mach 44(4):217–222
22. Tsoulias N, Paraforos DS, Xanthopoulos G et al (2020) Apple shape detection based on geometric and radiometric features using a LiDAR laser scanner [J]. Remote Sens 12(15):2481
23. Sun Sashuang, Wu Q, Jianchang Tan et al (2015) Research on target recognition and reconstruction of single apple under branch occlusion [J]. J Northwest A&F University (Natural Science Edition) 45(11):138–146
24. Feng J, Zeng L, He L (2019) Apple fruit recognition algorithm based on multi-spectral dynamic image analysis [J]. Sensors 19(4):949
25. Zhao J, Tow J, Katupitiya J (2005) On-tree fruit recognition using texture properties and color data [C]. International Conference on Intelligent Robots and Systems, pp 263–268
26. Wu G, Li B, Zhu Q et al (2020) Using color and 3D geometry features to segment fruit point cloud and improve fruit recognition accuracy. Comput Electron Agric 174:105475
27. Liu X, Zhao D, Jia W, Ji W, Sun Y (2019) A detection method for apple fruits based on color and shape features. IEEE Access 7:67923–67933
28. Linker R, Cohen O, Naor A (2012) Determination of the number of green apples in RGB images recorded in orchards. Comput Electron Agric 81:45–57
29. Liu Lijuan, Dou Peipei, Shine Wong (2021) Study on the recognition method of overlapping and blocking apple images in natural environment. China J Agric Chem 42(6):174–181
30. Jana S, Basak S, Parekh R (2017) Automatic fruit recognition from natural images using color and texture features [C]//2017 Devices for Integrated Circuit (DevIC). IEEE, pp 620–624
31. Xu L, Lv J (2018) Recognition method for apple fruit based on SUSAN and PCNN [J]. Multimedia Tools Appl 77(6):7205–7219
32. Gongal A, Silwal A, Amatya S et al (2016) Apple crop-load estimation with over-the-row machine vision system [J]. Comput Electron Agric 120:26–35
33. Si Yongsheng, Liu Gang, Gao Rui (2009) Green Apple identification technology based on K-means clustering [J]. J Agric Mach 40(S1):100–104
34. Kurtulmus F, Lee WS, Vardar A (2011) Green citrus detection using 'Eigen Fruit', color and circular Gabor texture features under natural outdoor conditions [J]. Elsevier Science Publishers B.V., pp 140–149
35. Rakun J, Stajnko D, Zazula D (2011) Detecting fruits in natural scenes by using spatial-frequency based texture analysis and multiview geometry. Comput Electron Agric 76(1):80–88
36. Pothen ZS, Nuske S (2016) Texture-based fruit detection via images using the smooth patterns on the fruit [C]. IEEE International Conference on Robotics and Automation (ICRA), pp 5171–5176
37. Liao Wei, Zheng Lihua, Li Minzan, Sun Hong, Yang Wei (2017) Identification of green apples under natural illumination based on random forest algorithm. J Agric Mach 48(S1):86–91
38. Tao YT, Zhou J (2017) Automatic apple recognition based on the fusion of color and 3D feature for robotic fruit picking [J]. Comput Electron Agric 142:388–396

39. Jia W, Mou S, Wang J et al (2020) Fruit recognition based on pulse coupled neural network and genetic Elman algorithm application in apple harvesting robot [J] Int J Adv Robotic Syst 17(1):255791245
40. Li Nannan (2022) Research on apple target recognition algorithm based on lightweight deep learning network [D]. Northwest A&F University.
41. Li Linsheng, Zeng Pingping (2019) Apple target detection based on improved Faster-RCNN framework of deep learning [J]. Mach Des Res 35(5):24–27. (in Chinesewith English abstract)
42. Chu PY, Li ZJ, Lammers K et al (2021) Deep learning-based apple detection using a suppression mask R-CNN [J]. Pattern Recognit Lett 147:206–211
43. Jing Weibin, Li Cunjun, Jing Xia et al (2019) Apple tree side view fruit recognition based on deep learning [J]. China Agric Inf 31(5):75–83
44. Gao F, Fu L, Zhang X, Majeed Y, Zhang Q (2020) Multiclass fruit-on-plant detection for apple in snap system using Faster R-CNN. Comput Electron Agric 176:105634
45. Jia W, Tian Y, Luo R et al (2020) Detection and segmentation of overlapped fruits based on optimized mask R-CNN application in apple harvesting robot [J]. Comput Electron Agric 172:105380
46. Wang D, He D (2022) Fusion of mask RCNN and attention mechanism for instance segmentation of apples under complex background [J]. Comput Electron Agric 196:106864
47. Jia W, Zhang Z, Shao W et al (2022) RS-Net: Robust segmentation of green overlapped apples [J]. Precision Agric 23(2):492–513
48. Chen W, Zhang J, Guo B et al (2021) An apple detection method based on Des-YOLOv4 algorithm for harvesting robots in complex environment [J]. Math Prob Eng 2021:1–12
49. Mao Tengyue, Song Yang, Zheng Lu (2022) Apple target detection based on multi-scale and mixed attention mechanism. Journal of South-Central University for Nationalities (Natural Science Edition), 41(2):235–242
50. Ma Z, Li NQ (2022) Improving apple detection using RetinaNet [J]. Lect Notes Electr Eng 813:131–141
51. Yue Linqian, Li Wenkuan, Yang Xiaofeng et al (2022) Apple detection and fruit diameter estimation method based on improved YOLOv4. Laser J 43(2):58–65. https://doi.org/10.14016/j.cnki.jgz2022.02.058
52. Wu Xing, Qi Zeyu, Wang Longjun et al (2020) Apple detection method based on lightweight YOLOv3 convolutional neural network. J Agric Mach 51(8):17–25
53. Gao Fangfang, Wu Zhenchao, Suorui et al (2021) Apple detection and video counting method based on deep learning and target tracking. J Agric Eng 37(21):217–224
54. Lu S, Wen K, Xin Z, Manoj K (2022) Canopy-attention-YOLOv4-based immature/mature apple fruit detection on dense-foliage tree architectures for early crop load estimation. Comput Electron Agric 193:106696
55. Zhao Dean, Wu Rendi, Liu Xiaoyang et al (2019) Apple positioning based on YOLO deep convolutional neural network for picking robot in complex background [J]. Trans Chin Soc Agric Eng (Trans of the CSAE) 35(3):164–173. (in Chinese with English abstract)
56. Tian Y, Yang G, Wang Z et al (2019) Apple detection during different growth stages in orchards using the improved YOLO-V3 model [J]. Computers Electron Agric 157:417–426
57. Zhao Hui, Qiao Yanjun, Wang Hongjun et al (2021) Apple fruit recognition in complex orchard environment based on improved YOLOv3 [J]. Trans Chin Soc Agr Eng (Transactions of the CSAE) 37(16):127–135 (in Chinese with English abstract)
58. Lv J, Xu H, Han Y et al (2022) A visual identification method for the apple growth forms in the orchard [J]. Comput Electron Agric 197:106954
59. Yan B, Fan P, Lei X et al (2021) A real-time apple targets detection method for picking robot based on improved YOLOv5 [J]. Remote Sens 13(9):1619

60. Yan Bin, Fan Pan, Wang Meirong et al (2022) Real-time apple picking pattern recognition for picking robot based on improved YOLOv5m [J] Trans Chin Society for Agric Mach 53(9):28–38, 59 (in Chinese with English abstract)
61. Sun Jun, Qian Lei, Zhu Weidong et al (2022) Apple detection in complex orchard environment based on improved RetinaNet [J]. J Agric Eng 38(15):314–322
62. Li H, Guo W, Lu G et al (2022) Augmentation method for high intra-class variation data in apple detection [J]. Sensors 22(17):6325

Chapter 9
UAV-Based Apple Flowers Pollination System

Shahram Hamza Manzoor, Muhammad Hilal Kabir, and Zhao Zhang

Abstract This book chapter delves into the realm of Unmanned Aerial Vehicles (UAVs) and their potential impact on apple flower pollination, offering a captivating exploration of this subject matter. Utilizing contemporary research and practical implementations, we explore the advancements of Unmanned Aerial Vehicle (UAV) technology and its potential to augment orchard productivity. As we explore the core of the subject, we reveal the significant significance of adequate pollination for apple trees. This study investigates the consequences of inadequate pollination, encompassing diminished fruit production and yield, irregular patterns of fruit-bearing, and compromised overall tree vitality. This establishes the foundation for the novel application of Unmanned Aerial Vehicles (UAVs) as airborne agents for pollination.

Expanding our investigation, we proceed to examine the different methods of pollination and the conventional methodologies and technologies utilized in pollination. This book chapter then discusses the underlying mechanisms of unmanned aerial vehicle (UAV)-based pollination, encompassing a range of topics including autonomous flight patterns and accurate pollen delivery. By conducting a comparative analysis between unmanned aerial vehicles (UAVs) and conventional pollinators, significant insights can be obtained regarding the potential advantages and obstacles associated with the integration of this technological innovation.

In short, this chapter functions as a symbol of excitement regarding the prospective development of apple orchards.

Keywords Unmanned Aerial Vehicle (UAV) · Pollination

S. H. Manzoor · M. H. Kabir · Z. Zhang (✉)
Key Laboratory of Smart Agriculture System Integration, Ministry of Education, China Agricultural University, Beijing 100083, China
e-mail: zhaozhangcau@cau.edu.cn

Key Lab of Agricultural Information Acquisition Technology, Ministry of Agriculture and Rural Affairs, China Agricultural University, Beijing 100083, China

College of Information and Electrical Engineering, China Agricultural University, Beijing 100083, China

M. H. Kabir
e-mail: mkhilal@zju.edu.cn

Z. Zhang and X. Wang (eds.), *Towards Unmanned Apple Orchard Production Cycle*, Smart Agriculture 6, https://doi.org/10.1007/978-981-99-6124-5_9

9.1 Introduction

The crucial process in apples production requires pollination whereby pollen is transferred from the male reproductive structure (stamen) to the female reproductive structure (pistil) as shown (Fig. 9.1) . This ensures that fruit is developed within an apple blossom [1, 2]. The success of this mechanism cannot be over-emphasized as it transforms fertilized pistils into matured apples. Pollination results when there is anthropogenic share (human intervention), wind, or insect-mediated activities in orchards since these guarantee adequate cross-pollination happens. Ultimately if accomplished successfully, apples would grow and become ready for harvesting throughout autumn right off an apple tree [1].

Pollination plays a crucial role in nature's reproductive cycles particularly so for apple trees along with many other plant species making it essential for ensuring genetic diversity within ecosystems. While habitat destructions caused by physical intervention along with chemical intervention such as pesticide use continue further reducing numbers [3, 4]; hence devising alternative pollination methods such as unmanned aerial vehicles (UAVs) gains significance. Drone technology has seen exponential growth and depicts potential advantages for various agriculture applications, including facilitating successful pollination procedures. Specifically enhanced efficacy, reduced labor requirements, and minimized risk of harm from chemicals commonly used in other agricultural practices present UAVs as a viable candidate over traditional means. Additionally, using drones can alleviate further caveats like harsh weather conditions and lack of access to experts. Several studies in recent times have shown promising results supporting interest in drone-based apple-pollination efforts.

9.2 Importance of Proper Pollination for Apple Trees

Several studies have demonstrated the significance of adequate pollination for the growth and productivity of apple trees. Insufficient pollination can result in notable statistical impacts on many aspects of apple blossom growth and the overall well-being of the tree. As per the findings of [5], inadequate pollinator activity or insufficient pollen transfer resulted in fruit set of less than 5%, indicating a considerable number of flowers that failed to develop into fruits [6]. According to Bradford's (2011) findings, insufficient pollination led to the production of apples that were deformed, undersized, and had suboptimal coloration, low sugar levels, and diminished firmness [7]. Klein (2019) emphasized the significance of insufficient pollination on yield, revealing a potential decrease of up to 50% in apple tree yield due to poor pollination as opposed to adequate pollination [8]. Furthermore, Kacira and Kacira (2010) have illustrated that inadequate pollination has a disruptive effect on the typical bearing cycle of apple trees, resulting in irregular fruit yield [9]. Van der Steeg (2017) established a correlation between deficient pollination and diminished

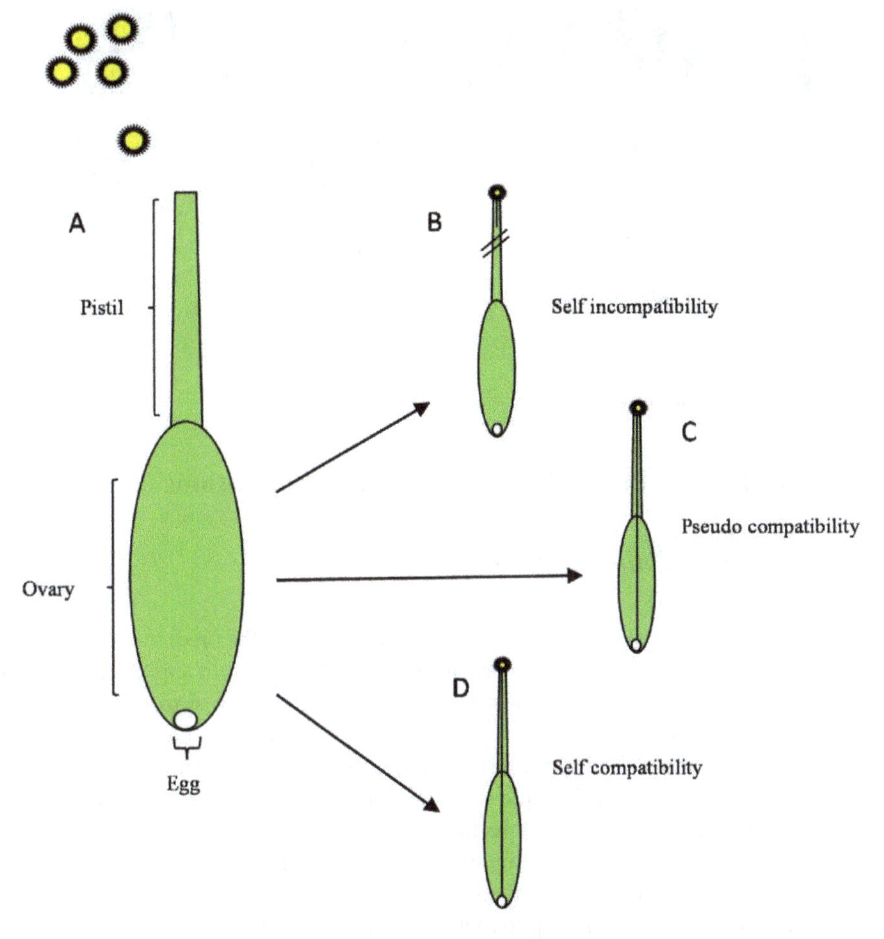

Fig. 9.1 illustrates the interactions and compatibility of pollen. In scenario A, the pollen is located in close proximity to the stigma. In scenario B, the pollen successfully lands on the stigma but fails to undergo germination. In scenarios C and D, the pollen is capable of germinating and successfully travels to the ovary [1] *Source* (GrowVeg.com)

tree vitality, as evidenced by heightened pest and disease susceptibility in apple trees. The statistical results underscore the crucial significance of employing efficient pollination management techniques to enhance fruit set, guarantee superior fruit quality, optimize yield, encourage consistent bearing, and sustain the general health and vigor of apple trees.

Fig. 9.2 Natural apple flower pollination via bee. *Source* (GrowVeg.com)

9.3 Traditional Pollination Methods and Limitations

9.3.1 Natural Pollination Methods

Bees, such as honeybees, bumblebees, and solitary bees, are the predominant natural pollinators of apple trees [10]. The sweet nectar and vibrant hues of apple tree flowers serve as attractants for bees as shown in Fig. 9.2. As they traverse from one flower to another, they facilitate the transfer of pollen from the stamen to the pistil, thereby enabling fertilization and subsequent fruit maturation. Additional pollinators of apple trees comprise of insects such as flies, beetles, and moths. However, their efficiency is lower than that of bees, therefore they may not provide enough pollination if used alone. Natural apple tree pollination relies on the existence of pollinators. As weather patterns, and the time of the bloom, are changing due to global warming, this is resulting in reduction of popuplation of natural pollinators. Therefore UAV based pollination could be a best solution for apple flower pollination [11].

9.3.2 Hand Pollination

Conventional methods of agricultural pollination include hand-pollinating flowers with either brushes, cotton swabs or some small eletronic pollination equipments as show in (Fig 9.3) in order to move pollen between stamen and pistil effectively. Growers typically resort to this approach when reducing bee populations threaten yields or require specific supervision regarding crop development timelines. By controlling how quickly and consistently flowers receive pollen through hand-pollinated methods, farmers might realize more productive outcomes. However, this technique presents challenges related to staffing needs due it is extensive labor requirements which could limit implementation on larger-scale farming operations.

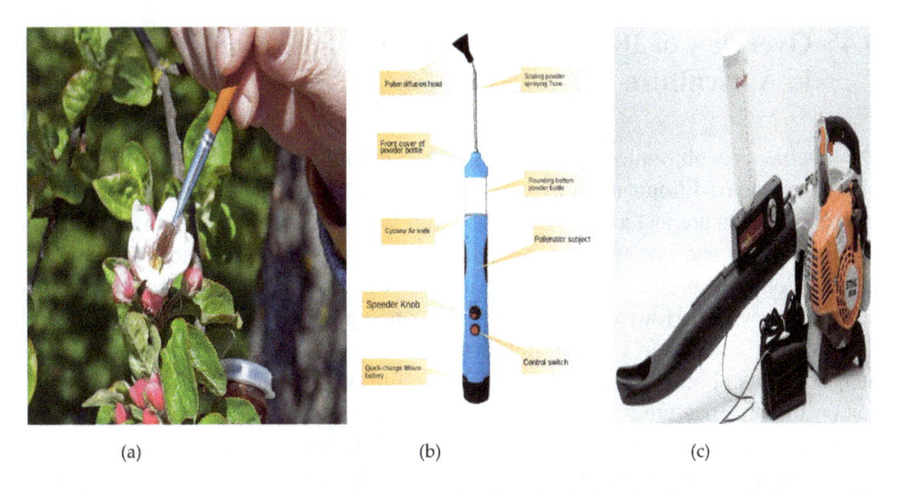

Fig. 9.3 Illustrates some conventional methods of pollination by hand; **a** Hand apple flower pollination. *Source* (gardeningknowhow.com). **b** and **c** Electric Pollen dispensors. *Source* Amazon.com

Moreover, such methods might result in imbalances within fruit growth caused by difficulties in ensuring the even spread of pollen across all flowers. Hense to reduce labor costs, and for timely and precise pollination UAV based pollination offers more and better benefits [12].

9.3.3 Limitations of Traditional Pollination Methods

Apple growers face numerous challenges with natural pollination processes that cannot be ignored for long-term sustainability. The decline in pollinator populations is a central issue stemming from habitat loss, pesticide usage, and climate change effects [3, 4]. Apple orchards are witnessing lower yields due to poor quality fruits causing major setbacks in their business strategies as they heavily depend on these products for their sales revenue [10]. Traditional techniques requiring specific weather conditions for optimal outcomes hinder orchard productivity; colder seasons or dryness reduce chances of successful fruit set during flowering periods while hotter days affect heating non-native flower parts reducing pollen viability leading towards lower productivity rates which affects fruits yield negatively as well. Manually performed conventional techniques impose significant labor costs since trained experts should perform this task carefully; untrained labor leading to inconsistent results reducing product value causing trouble sorting out varieties sizes. There has been a rise in interest in using drones for pollination as a result of the limitations of conventional approaches. In the following paragraphs, we'll look into whether or not drones could be used as a viable alternative pollination method in apple orchards.

9.4 Overview of Drone Technology and Potential Use in Agriculture

Drones have revolutionized many sectors by providing low-priced and high-yield alternatives to traditional methods. Crop mapping, plant health monitoring, and pesticide application are just a few of the ways in which drones might aid the agricultural industry. However, one of the most exciting uses of drones in agriculture is pollination. Pollinators, such as bees, are critical to the development of numerous crops around the world. However, the diminishing bee population as a result of several causes such as climate change, habitat degradation, and pesticide use has generated concerns about agricultural pollination's future. As a result, employing drones for pollination has emerged as a potential answer. Drone technological advancements have enabled drones to be programmed to conduct precise motions and deliver payloads properly, making them perfect for pollination. Drones can more efficiently distribute pollen across crops than traditional hand pollination methods. Using unmanned aerial vehicles (UAVs) to pollinate apple trees could have a lot of benefits. Using drones to spread pollen between apple flowers could improve pollination rates, leading to more fruit set and a higher yield. This new method uses new technology to deal with problems that might come up because the number of bees is going down. This makes sure that apple orchards will be productive and able to keep going in the future.

9.4.1 Advantages and Challenges Associated with Drone Pollination

The implementation of drones for plant pollination has numerous favorable consequences. Notably some types of drones are capable of operating seamlessly under various weather conditions including strong winds and rain showers. This characteristic enables non-stop pollination activities throughout the day. Ultimately enhancing overall effectiveness considerably. Moreover. Drones possess adeptness at covering expansive areas swiftly—fostering efficient and uniform dispersion of pollen. Furthermore. Relying on drones for plant pollination reduces reliance on labor-intensive manual methods. Offering a cost-effective alternative to hand pollination techniques. The subsequent reduction in crop production costs not only benefits farmers but proves especially advantageous for those residing in developing countries as it contributes towards their economic sustainability. However. There are difficulties associated with using drones for pollination purposes as well. There exists a crucial challenge surrounding the limited carrying capacity inherent to these machines. Pollen carries significant mass which necessitates larger motors and batteries within drone construction to accommodate its transport. To ensure uniform distribution across target regions it becomes vital that these unmanned aerial vehicles scatter pollen meticulously and uniformly. Lastly. In order to accurately identify and locate flowers within the desired area. Drones must be equipped with high-resolution cameras or similar sensors. This requirement plays a pivotal role in ensuring successful pollination outcomes.

9.4.2 *Methods of Drone Pollination*

Drone pollination is a unique method of pollination in which drones are used as a substitute for or in addition to conventional pollinators like bees. Drone pollination can be done either by dry pollination or wet pollination.

9.4.2.1 Dry Pollination

In dry pollination, pollen is sprayed onto the desired flowers using a drone. The pollen is sprayed evenly onto the blooms when the drone flies over the designated region. The male flowers' dry pollen is gathered in a receptacle on the drone until it is time to be dispersed. The effectiveness of dry pollination is one of its key benefits. Compared to manual pollination, this method is far more efficient and can cover more ground in less time. However, dry pollination has the potential drawback of having the pollen blown away from where it's needed. This may lead to less efficient pollination.

9.4.2.2 Wet Pollination

To perform wet pollination, pollen is diluted in a liquid and then sprayed onto the desired blossoms. The liquid solution, which is typically a combination of water, sugar, and nutrients, aids pollen in sticking to the flowers as shown in Fig. 9.4(b). If a better pollination rate, wet pollination is the way to go. The liquid solution not only nourishes the blooms, but also shields the pollen from the wind, resulting in stronger, healthier fruit. However, moist pollination is more difficult and requires more sophisticated tools than dry pollination.

9.4.2.3 Direct Pollination

Drones can be used for direct pollination, in which pollen is spread directly onto flower stigmas. A pollen applicator, or a specially designed brush or wand, can be attached to the drone to spread the pollen. The drone is flown over the orchard, with certain trees or regions in mind, and the pollen is dropped off while hovering over the blossoms. The high success rate of fruit set is guaranteed by direct pollination's tailored pollen distribution. Apples, for example, benefit greatly from cross-pollination between different apple kinds, and this method is ideal for pollinating such

(a) (b)

Fig. 9.4 **a**, and **b** Quad bike broadcast for Pollen application. *Source* kiwipollen.com

crops. However, this strategy may be constrained by the drone's payload capacity and calls for precise calibration of the pollen delivery device.

9.4.2.4 Indirect Pollination

With drone-based indirect pollination, a cloud of pollen is released over the orchard. The pollen can be released from the drone in the form of a fine mist that the wind can then carry to surrounding flowers. The accuracy of the pollen delivery mechanism is less crucial for indirect pollination, which benefits the orchard as a whole. However, because not all of the pollen reaches the intended flowers, fruit set may be lower than with direct pollination.

9.4.2.5 Combination of Direct and Indirect Pollination

The optimization of drone pollination efficacy can be achieved through the utilization of both direct and indirect pollination techniques. The utilization of drones can facilitate the direct delivery of pollen to the intended flowers located in the central region of the orchard, in addition to the dissemination of a pollen cloud over the adjacent trees. The aforementioned approach offers the benefit of accurate and focused transportation of pollen, as well as comprehensive dissemination throughout the entire orchard. Nonetheless, precise synchronization and adjustment of the drone's mechanism for dispensing pollen are necessary.

9.4.3 Pollen Dispenser Development

In order to enhance the effectiveness of drone-mediated pollination, researchers are currently engaged in the development of advanced pollen dispensing mechanisms.

9.4.3.1 Electrostatically Charged Pollen

A technique being explored to increase the efficiency of drone pollination is through the application of electrostatically charged pollen. Adding an electric charge to the pollen grains makes them more sticky, ensuring they stick to the blooms. It is hypothesized that increased pollen transfer and fruit set could come from using positively charged pollen that is attracted to the negatively charged surface of the bloom. Using pollen that has been electrostatically charged has a number of benefits. For instance, it is anticipated that the charged pollen particles will be more likely to cling to the designated target flowers, hence increasing the accuracy and efficiency of pollen transmission. More fruit set and harvest may ensue if more pollen is carried to the flowers. However, there are also certain difficulties connected with employing pollen that has an electrical charge. Overcharging pollen might lead it to clump together, reducing its efficacy, so care must be used when charging it. Drone pollination may be more expensive and difficult to implement because charging may necessitate specialist equipment and expertise.

9.4.3.2 Gel-Based Pollen Carriers

Another approach that is currently under development to enhance the efficiency of drone pollination involves the utilization of pollen carriers that are based on gel. The procedure involves mixing the pollen with a viscous polymer that shields it from environmental factors like precipitation and air currents while also improving its transport efficiency. There are many advantages to using gels as pollen carriers in the context of drone pollination. The establishment of a protective envelope surrounding the pollen grains, for example, aids in decreasing the impacts of wind dispersal, boosting the efficiency of pollination and the frequency with which fruit is produced. Nutrients that may improve the fruit's overall health and growth could be incorporated into the gel matrix. However, there are additional drawbacks associated with using gels as pollen transporters. To ensure the pollen and flowers remain healthy, the formulation of the carrier substance must be carefully considered. Drone pollination processes are already complex, and they may be more difficult to manage when using pollen that comes in a gel form.

9.5 Literature Review

[13] This paper introduces an innovative approach for detecting apple flowers, which relies on characteristics derived from a convolutional neural network (CNN) feature extraction as illustrated in Fig. 9.5. The objective is to implement more precise and efficient methodologies for the assessment of bloom intensity, a crucial factor in enhancing fruit yield. Conventional techniques that rely on basic color thresholding encounter obstacles due to the presence of inconsistent lighting circumstances and obstructions caused by leaves, stems, or other flowers. The CNN-based approach that has been proposed exhibits encouraging outcomes in precisely identifying apple flowers and approximating bloom intensity.

[12] Walnut trees can be affected by a lack of pollination for a number of reasons, and this study investigates the possibility of creating an aerial artificial pollination system for these trees as shown in Fig. 9.6. Using a quadrotor UAV, the system releases pollen over the designated trees, with the use of dynamic modeling and computational fluid dynamic (CFD) simulations to guarantee its successful arrival. The research reveals details about the system's development and testing, as well as its possible agricultural uses.

[14] An effective and novel spray technique for artificial pollination of date palm using drones is shown in this research. An adequate number of fruit formed when the technique was applied to three different varieties. Since this method of pollination for date palms is more efficient and requires less human intervention, it has the potential to replace the conventional approach. However, more study is required to determine the best time and length of application, which may vary by cultivar and location. The next stage is to establish a framework for a commercially viable, autonomous electronic pollination program utilizing drones in date palm plantations throughout all producing regions as shown in Fig. 9.7.

[15] This study analyzes the effects of mechanical or chemical thinning on crop load, fruit size, and quality components such as colour, firmness, and sugar and acid content. The study identifies two groups of quality components: Group 1 characteristics include size, colour, skin performance, firmness and sugar and acid content of

Fig. 9.5 Images from the extra datasets AppleB (left), AppleC (center), and Peach (right), overlaid with detections acquired by this approach [13]

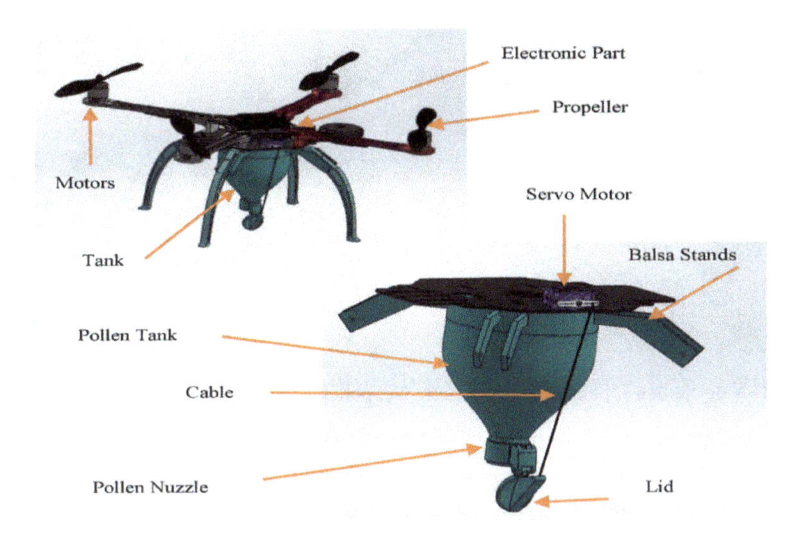

Fig. 9.6 Description of the quadrotor design [12]

Fig. 9.7 Date palm pollination methods include traditional hand pollination **A** hand spray pollination **B** and drone pollination **C** [14]

the fruit. Group 2 characteristics were represented by inorganic components, especially calcium and potassium which are implicated in the susceptibility of fruit to physiological disorders. While group 1 characteristics were improved by increasing thinning intensity, storability of the fruit was better at high than at low crop loads. Therefore, a compromise between all quality requirements must be found for a good economic return.

[16] The study demonstrates the feasibility of using UAVs with RGB cameras for quantifying pear flower clusters. This study aligns with the trend of using UAVs in agriculture, including for apple orchards, where UAVs have been explored for flower pollination. Apple orchards heavily rely on pollination, and the use of UAVs can improve pollination efficiency and increase crop yield. By combining UAV

Fig. 9.8 A dry pollen application system (left), Wet pollen spray application system (right) [17]

technology with advanced imaging techniques and image processing algorithms, researchers can continue to develop innovative solutions for optimizing apple orchard management and improving crop productivity.

[17] This article discusses the development of a robotic kiwifruit pollinating system that uses machine vision and a convolutional neural network to detect and localize kiwifruit flowers as depicted in Fig. 9.8. While the current system is not yet commercially viable, future work will focus on scaling up the system to accommodate a full orchard row width of 4.5 m and optimizing the machine vision software for improved shot accuracy.

[11] The study explores the feasibility of using UAVs for apple flower pollination as shown in Fig. 9.9. This application of UAV technology aligns with the growing trend of using UAVs in agriculture to monitor crop growth, health, and yield potential. Apple orchards heavily rely on pollination, and the use of UAVs for this purpose is a new, emerging area of research. The study found that UAVs equipped with specialized equipment, such as brushes or sprayers, can be used to distribute pollen effectively and efficiently. By using UAVs, farmers can potentially save labor costs, improve pollination success, and increase crop yield. This study contributes to the emerging area of research on the use of UAVs in agriculture and highlights the potential benefits of using UAVs for apple flower pollination.

[18] Apple flower pollination with UAVs is a potential application of the study of how to effectively classify flowers using machine learning techniques. The proposed method begins with segmenting flower photos to get rid of complicated backgrounds, then moves on to feature extraction using SIFT and SFTA techniques, and finally concludes with classification using SVM and RF algorithms. Experiments performed with a dataset containing 215 photos of flowers demonstrated that SVM-based algorithms achieved higher accuracy than RF algorithms employing SIFT or SFTA as feature extraction algorithms. The efficiency and accuracy of UAV-based pollination could be increased by adapting this method for the identification of apple blooms.

[19] The elimination of insect pests is a prerequisite for long-term agricultural success. The widespread use of pesticides to combat these nuisances, however, raises

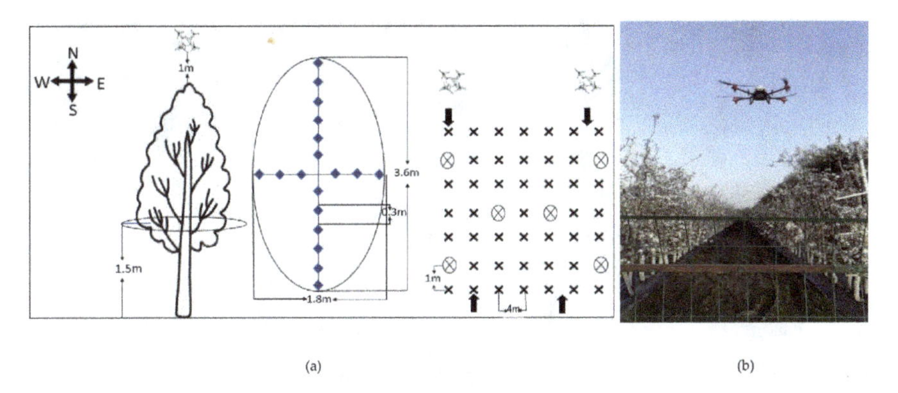

Fig. 9.9 **a** Illustration of water-sensitive paper sampling spots and UAV flight path. **b**. Field testing of UAV pollination [11]

environmental and public health problems. Spot spraying, which helps reduce pesticide use through early detection of insect pests, can be implemented. This means that insect pest detection and classification systems are becoming increasingly important. To identify and categorize insect pests in apple blooms using UAV footage, this research offers an object identification system based on YOLO object detection architectures. With the suggested approach, insect pests can be detected and identified with a high degree of precision, allowing farmers to use fewer pesticides while improving productivity and sustainability in apple orchards as depicted in Fig. 9.10.

[20] The utilization of Unmanned Aerial Vehicles (UAVs) in the agricultural sector has resulted in an increase in the agricultural UAV market. However, Nano Aerial Vehicles (NAVs) remain under utilized in this domain. Navigation Autonomous Vehicles (NAVs) possess a distinct advantage over conventional Unmanned Aerial Vehicles (UAVs) in terms of precision and accuracy in task execution, owing to their unique physical attributes. This makes them a suitable choice for carrying out precision agriculture. In order to advance the study of NAVs in the field of agriculture, a novel open-source solution, referred to as Nano Aerial Bee (NAB), was devised with the aim of emulating and supporting bees in the process of pollination. The intricacy of this prototype is attributed to the intricate interplay among various hardware components and the requirement for self-sufficient flight that can perform pollination. Apart from the NAB solution, three distinct iterations of YOLO were trained on flower detection datasets to aid in identifying the spatial orientation of apple flowers within an image, thereby defining the NAB execute command. The performance of the models trained on the Flower Detection Dataset is deemed satisfactory, with YOLOv7 and YOLOR exhibiting the most optimal outcomes. The aforementioned advancements exhibit auspicious ramifications for the application of Unmanned Aerial Vehicles (UAVs) in the pollination of apple blossoms (Fig. 9.11).

[21] UAVs have been widely used in agriculture, including for apple orchard management, where the thinning process is essential to improve fruit quality as shown in Fig. 9.12. Traditional methods rely on human visual inspection of flowering

Fig. 9.10 Detection results of YOLOv5x [19]

(a) YOLOv5 (b) YOLOv7 (c) YOLOR

Fig. 9.11 Flower detection in samples from the tensor flow flower detection test set [20]

Fig. 9.12 Examples of view points [21]

intensity, but recent research has explored the use of ground-level images to achieve better results. This study investigates the potential of UAV high-resolution RGB imagery to measure flowering intensity by segmenting the white pixels corresponding to apple flowers. However, the results show low correlation between the percentage of white pixels per tree and flowering intensity, indicating the complexity of working with drone images. Nevertheless, further exploration of alternative approaches should be considered before discarding the use of UAV RGB imagery for estimation of flowering intensity and ultimately for pollination purposes in apple orchards.

[22] This paper presents an algorithm to detect and count yellow tomato flowers in a greenhouse using a drone with adaptive global threshold, segmentation over the HSV color space, and morphological cues. The algorithm can be used for apple flower pollination and yield estimation.

[23] The establishment of a technique for identifying and identifying flowers through the utilization of color images captured by an unmanned aerial system (UAV) can furnish valuable insights for the surveillance of flower growth in agricultural fields. The present investigation centered on cotton blooms and involved the training of a convolutional neural network (CNN) to identify the blooms in unprocessed unmanned aerial vehicle (UAV) images. The 3D coordinates of the blooms were subsequently determined by means of the dense point cloud generated via the structure from motion technique. A constrained clustering algorithm was developed to register the same bloom detected from different images based on the 3D location of the bloom. The proposed method provides a high-throughput approach to continuously monitor the flowering progress of cotton, which can be adapted for other crops such as apple flowers. However, limitations of the UAV imaging system can affect the accuracy of the bloom registration result, such as hidden blooms that were not captured by the aerial images, which can underestimate the number of blooms. Nevertheless, the use of UAV technology in flower monitoring can aid in production management, yield estimation, and selection of specific genotypes for crop improvement (Fig. 9.13).

[24] Regular monitoring and assessment of crops is crucial for optimal crop production. For fruit trees, such as apple trees, the bloom period is important as

Fig. 9.13 The bloom detection results for plot 0110 and plot 1011 in field 1 on 8/12/2016 dataset. The left images show the point cloud and detected blooms and right images show the corresponding blooms in the images [23]

it correlates with the fruiting process. This study developed an image processing algorithm to detect peach blossoms on trees using aerial images from an off-the-shelf unmanned aerial system (UAS) equipped with a multispectral camera. The algorithm involved contrast stretching and thresholding segmentation to detect the peach blossoms. The study demonstrated the potential of using UAS and image processing algorithms as a monitoring tool for orchard management, which can also be applied for apple flower pollination (Fig. 9.14).

[25] This paper describes the development of a small drone's visual navigation capability, a critical step toward creating a fully autonomous nanodrone for pollinating flowers. To detect and approach flowers, the drone employs a color camera and a ToF distance sensor. The flower is located using object detection CNN and then approached and touched using a two-stage visual serveying algorithm. The drone successfully detects, approaches, and touches flowers autonomously after being trained on an artificial dataset, with potential applications for pollinating various flower types, including apple flowers (Fig. 9.15).

[26] The present study presents a novel automated approach employing a residual convolutional neural network (CNN) for the estimation of bloom intensity in fruit production. Through the process of fine-tuning a Convolutional Neural Network (CNN) using apple flower images, the network is able to achieve a high level of accuracy in identifying individual instances of flowers across different species and environmental conditions. The method that has been proposed exhibits potential for the advancement of unmanned aerial vehicles (UAVs) in the context of apple flower pollination. It presents a dependable and effective strategy for the detection and recognition of flowers in unregulated environments (Fig. 9.16).

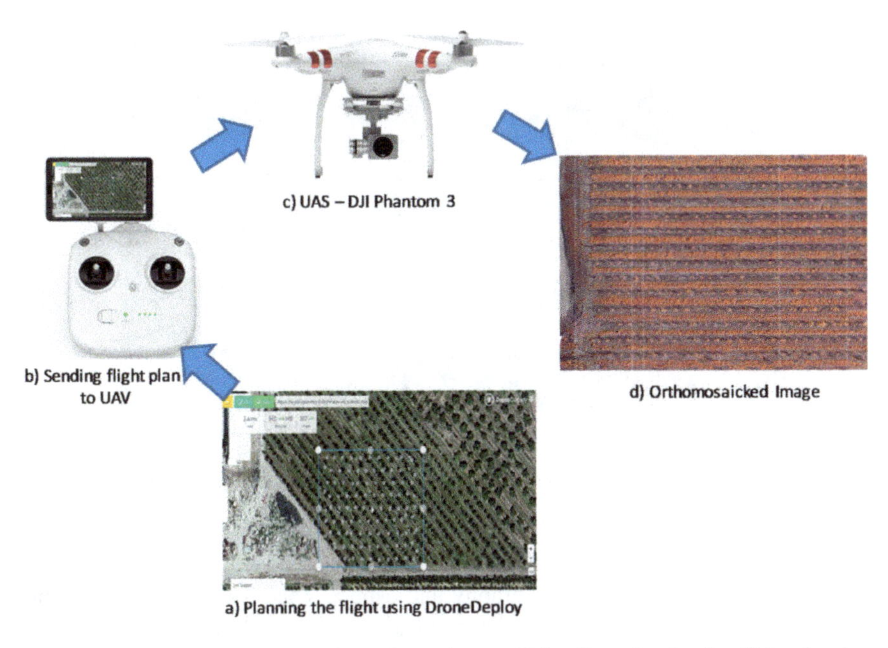

Fig. 9.14 Using DroneDeploy for UAV Operations. **a** Using DroneDeploy for flight planning; **b** transmitting the flight plan to the UAS; **c** the DJI Phantom 3 UAS; and **d** the resulting orthomosaicked image [24]

Fig. 9.15 (Columnwise) Successful navigation trials towards the pollination position [25]

[27] The present study investigates the apple yield and quality in a 0.8-hectare orchard located in northern Greece, with a particular focus on site-specific management strategies. The study focused on the examination of two apple cultivars, namely Red Chief and Fuji, throughout a span of two consecutive growing seasons. The research findings demonstrate a robust association between the quantity of flowers and the yield of crops, facilitating accurate thinning practices and early estimation

Fig. 9.16 Illustrations of flower detection within a single image of the Pear dataset [26]

of crop yield. The presence of spatial variations in apple yield and quality can be attributed to the influence of topography and aspect. The results indicate possible enhancements in the utilization of unmanned aerial vehicles (UAVs) for the purpose of apple flower pollination, thereby augmenting both the quantity and quality of crops within a specific geographical setting [28]. This research investigates the viability of utilizing apple flowering distribution maps as a means of implementing precise orchard management strategies. The assessment of flower density variability was carried out in a commercial apple orchard located in Central Greece, utilizing image analysis techniques. A computational algorithm was devised with the purpose of forecasting tree yield by utilizing captured images. The results indicate that the utilization of flowering distribution maps can facilitate early yield forecasting, thereby assisting in the effective management of orchards during the entire growth period. By employing this dataset, the utilization of unmanned aerial vehicles (UAVs) for the purpose of apple flower pollination can be advantageous due to the ability to make more informed management decisions based on location as shown in Fig. 9.17.

[29] The success of apple fruit production depends on individual tree physiological conditions and requires site-specific management practices. This study aimed to detect apple flowering abundance in a high-density apple orchard using image analysis of HSL images. The number of flower clusters (FC) per tree was estimated by optimizing the HSL thresholding algorithm. The study assumed three hypothetical tree-specific management practices based on FC thresholds of 25, 50, and 100, and assessed their performance. The images were acquired during the day and night using both a still camera and an industrial color camera. The results indicated that comparable FC counting performance was achieved using either a still camera or an industrial camera. When hypothetical spraying was carried out based on a 100 FC threshold using an industrial camera during the day, 10% incorrect executions were identified. This research could provide insights into potential site-specific management practices for apple orchards using UAVs for flower pollination (Fig. 9.18).

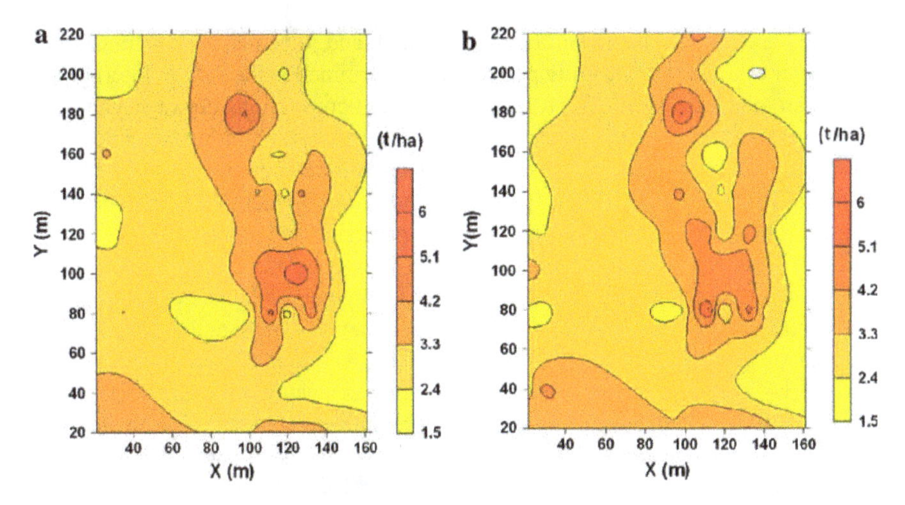

Fig. 9.17 Yield maps for measured yield **a** and anticipated yield **b** [28]

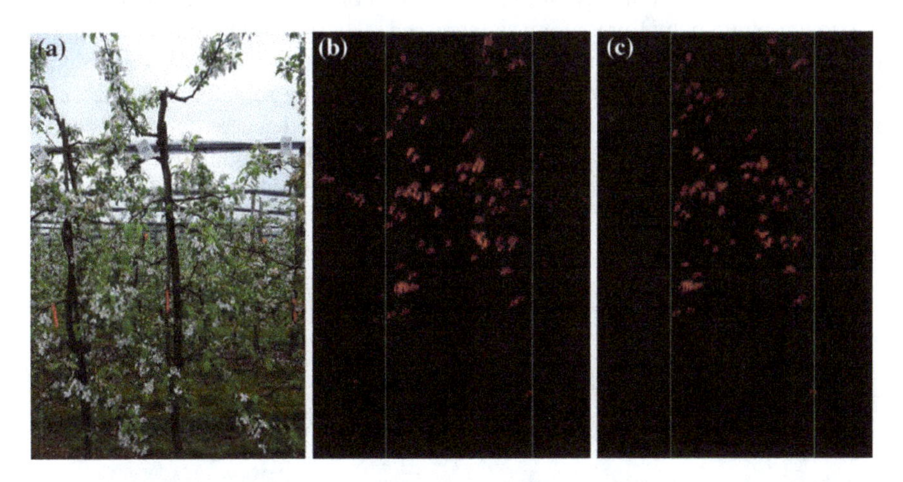

Fig. 9.18 Yield maps for (a) measurementImage processing example captured during the day with a static color camera. an Original image, b image after HSL thresholding, and c image after thresholding and rejection of things beyond the tree frame (green), as well as rejection of objects that are too large or too small. Figures depict a tree with 172 FC, whereas image analysis recognized 117 items (Color picture online) sured yield and (b) forecast yield. [29]

[30] In this study, precision agriculture tools were utilized to investigate the spatial variability of blooming, yield, and quality parameters in an apple orchard, with the aim of creating management zones to facilitate decision-making and practice evaluation. The orchard, located in the Research and Development Station for Pomology Voinesti, Dambovita, was 0.9 ha in size and consisted of Florina apple trees and Generos pollinators. GPS mapping was used to record the surface, and yield was determined by manual harvesting and weighing of each tree. Apple quality was

assessed based on firmness, soluble solid content, and juice pH, and the number of flowers per tree and the flowering period were determined in the spring. Interpolated maps were generated using Surfer 8 software, and correlations between yield and quality were established, with some quality characteristics exhibiting negative correlations with yield, such as firmness. Precision agricultural equipment, such as UAVs for apple flower pollination, can increase orchard management and fruit output by increasing the accuracy and efficiency of data gathering and processing.

[31] While liquid spray pollination (LSP) is widely used for pollinating fruit trees, little is known about how varying LSP parameters influence pollen activity. This research set out to examine three LSP parameters—the length of time pollen suspensions were stored, sprayer settings, and the direction of airflow from a UAVS's downwash—that affect pollen activity. The sprayer's settings included the rotary atomizer's (RA) speed, nozzle size, spraying pressure, and recirculation device type. Except for nozzle size and UAVS downwash airflow, the results showed that all other parameters significantly influenced pollen activity. Using the recirculation mechanism significantly reduced pollen activity in the tank, and storing pollen suspensions for longer periods of time also lowered activity. Pollen activity was reduced when maximum pump output pressure was increased, as was spraying pressure. Similar to how increasing the RA's rpm dampened pollen activity, so too did decreasing the RA's rpm. The data presented here can be used as a starting point for further research into improving pear LSP technology and maximizing pollen activity for successful pollination. In addition, integrating UAVS into LSP technology has the potential to provide a method for efficient, targeted pollination of apple blossoms (Fig. 9.19).

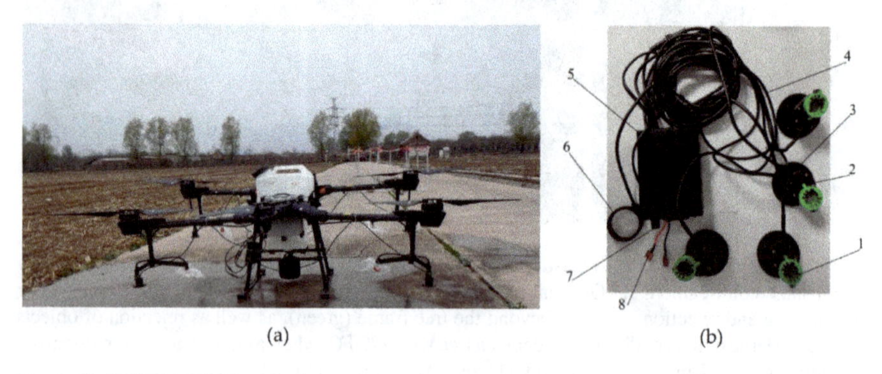

Fig. 9.19 DJI T20 UAVS with electronic RA system. (1) The nozzle body, (2) the brushless motor, (3) the fluted disc, (4) the power line, (5) the control box, (6) the induction ring, (7) the ESC, and (8) the electrodes. (a) DJI T20 UAVS with RA and (b) RA system controlled electronically [31]

9.6 Future Directions and Potential Applications of Drone Technology in Apple Pollination

9.6.1 Potential Applications of Drone Technology in Agriculture

Drones could change a lot of things about agriculture, including how apples are pollinated. There are several other ways that drone technology could be used in farmland besides pollination. For instance:

Crop monitoring: Crop monitoring can be achieved through the use of drones that are equipped with cameras and sensors. These devices can effectively detect issues such as pest infestations, disease outbreaks, and nutrient deficiencies, thereby ensuring the maintenance of crop health. The provision of this information can aid farmers in making better choices regarding fertilization, irrigation, and pest control (Fig. 9.20).

Precision farming: Farmers can use drones to make high-resolution images of their fields that show them where more or less fertilizer, water, or other inputs are needed. This can help cut down on waste and make crops grow better.

Pesticide application: Drones can be fitted with pesticide sprayers for more accurate and targeted spraying. This can lessen the pesticide's impact on non-target organisms and save money (Fig. 9.21)

Fig. 9.20 Crop monitoring via drone. *Source* (eos.com)

Fig. 9.21 Pesticide Spraying via drone. *Source* euractiv.com

9.6.2 Opportunities for Further Research and Development

Despite the significant advancements in drone technology in recent years, there remains considerable scope for further research and development. Several potential areas for improvement may be identified.

Battery life: The present battery life of drones is restricted, thereby constraining their operational range and efficacy. A longer duration of battery life would enable unmanned aerial vehicles to achieve prolonged flight durations and expand their coverage area.

Autonomous navigation: The concept of autonomous navigation entails the ability of drones to adapt to dynamic environmental conditions and effectively navigate intricate terrain, surpassing the mere capability of following predetermined flight paths that many drones currently possess.

Pollen dispensers: Researchers are working on new pollen dispensers that will increase the efficiency of drone pollination. Continued research in this field may result in even more effective and efficient pollination strategies.

9.6.3 Environmental and Ethical Considerations of Drone Pollination

Although drone pollination has many potential benefits, there are also environmental and ethical concerns to consider. Among the potential issues are:

Impact on wild pollinators: If drone pollination becomes common, it may lessen the demand for honeybees and other wild pollinators. While this may appear to be a beneficial thing at first, it may have unforeseen implications for the entire ecosystem. Some researchers believe that combining drones and natural pollination is the most effective strategy.

Energy use: Drones require energy to function, and this energy must come from someplace. Drone pollination may contribute to greenhouse gas emissions and climate change depending on how the energy is generated.

Privacy and safety: Drones can be obtrusive and perhaps harmful if they are not used safely. It will be critical to set legislation and guidelines to guarantee that drone pollination is done safely and responsibly.

Apple pollination is just one example of how drone technology might have a huge impact on agriculture. Drones will likely play an increasingly essential part in precision agriculture as they improve and become more proficient, assisting farmers to produce crops more efficiently and sustainably. To ensure that drone pollination is a sustainable and responsible practice, researchers must continue to investigate and address any environmental and ethical concerns.

9.7 Conclusion

9.7.1 Summary of Key Findings and Implications for Future Research and Practice

Drone pollination of apples is a promising technique that has the potential to change the agriculture economy. Several major findings emerged from our analysis of the literature:

Drones are effective pollinators: Studies have shown that drones can pollinate apple blooms effectively, with some even outperforming traditional honeybees.

Efficiency and precision: Drone pollination is more efficient and precise than traditional pollination methods, allowing for more effective resource usage and higher agricultural yields.

Room for improvement: While drone technology has advanced significantly, there is still much space for advancement in areas such as battery life, autonomous navigation, and pollen dispensers.

Environmental and ethical concerns: The use of drones in apple pollination poses a number of environmental and ethical concerns that must be addressed.

The findings have obvious implications for future study and practice. The broad use of drone pollination will require continued study and development in drone technology, particularly in areas like battery life and autonomous navigation. If drone pollination is to be a sustainable and responsible activity, more study of its effects on the environment and on people's values is required.

9.7.2 Recommendations for Apple Growers and Policymakers Regarding the Use of Drones in Apple Pollination

Following a comprehensive analysis of the existing literature, we propose a number of suggestions for apple cultivators and policy makers with regards to the implementation of unmanned aerial vehicles (UAVs) in the pollination of apple orchards.

Consider the potential benefits and drawbacks: Prior to implementing drone pollination, it is imperative for growers and policymakers to thoroughly evaluate the potential advantages and disadvantages, encompassing ecological and ethical factors.

Invest in research and development: It is recommended that apple growers allocate resources towards research and development endeavors pertaining to drone technology in order to enhance the efficiency and efficacy of pollination. Additionally, policymakers are encouraged to provide financial support for research initiatives in this domain.

Develop guidelines and regulations: In order to promote the secure and accountable deployment of drones in the context of apple pollination, it is recommended that policymakers devise a set of guidelines and regulations pertaining to drone operation, as well as establish optimal practices for growers.

Promote collaboration: Facilitating collaboration among researchers, growers, and policymakers is imperative for the efficacious implementation of drone pollination in apple orchards.

To sum up, the utilization of unmanned aerial vehicles in the process of apple pollination presents a hopeful substitute to conventional pollination techniques, exhibiting the capability to enhance efficacy, accuracy, and environmental soundness. Although there are still obstacles to be tackled, persistent investigation and cooperation among relevant parties can guarantee the safety, efficiency, and accountability of drone pollination.

References

1. Ramírez F, Davenport TL (2013) Apple pollination: a review. Sci Hortic (Am) 162:188–203. https://doi.org/10.1016/j.scienta.2013.08.007
2. Abbott DL (1970) The role of budscales in the morphogenisis and dormancy of the apple fruit bud. In: Luckwill LC, Cutting CV (eds) Physiology of tree crops. Academic Press, pp 65–82
3. Kline O, Joshi NK (2020) Mitigating the effects of habitat loss on solitary bees in agricultural ecosystems. Agriculture 10(4). doi: https://doi.org/10.3390/agriculture10040115
4. Obregon D, Guerrero OR, Stashenko E, Poveda K (2021) Natural habitat partially mitigates negative pesticide effects on tropical pollinator communities. Glob Ecol Conserv 28:e01668. https://doi.org/10.1016/j.gecco.2021.e01668
5. Sánchez-Estrada A, Vejarano R, García-López R, Hernández-Montiel LG (2018) Effect of pollination treatment on fruit set, fruit quality, and economic profitability in 'Golden Delicious' apple orchards. Spanish J Agric Res 16(3):e1008
6. Bradford KJ (2011) Pollen-pistil interactions and their role in regulating fruit set and development. J Exp Bot 62(1):62–63
7. Klein AM, Vaissière BE, Cane JH, Steffan-Dewenter I, Cunningham SA, Kremen C (2019) Importance of pollinators in changing landscapes for world crops. Proc R Soc B Biol Sci 286:20180624
8. Kacira M, Kacira A (2010) Effects of pollination on fruit set, growth and quality in apple. Turkish J Agric For 34(3):231–237
9. Van der Steeg J, Pree DJ, Blaauw BR, Isaacs R (2017) Pollination and pest management practices are positively associated with apple yield and fruit quality in Michigan orchards. Crop Prot 93:141–149
10. Pardo A, Borges PAV (2020) Worldwide importance of insect pollination in apple orchards: a review. Agric Ecosyst Environ 293:106839. https://doi.org/10.1016/j.agee.2020.106839
11. Wang Y et al (2022) Pollination parameter optimization and field verification of UAV-based pollination of 'Kuerle Xiangli'. Agronomy 12(10). doi:https://doi.org/10.3390/agronomy12102561
12. Mazinani M, Zarafshan P, Dehghani M, Vahdati K, Etezadi H (2023) Design and analysis of an aerial pollination system for walnut trees. Biosyst Eng 225:83–98. https://doi.org/10.1016/j.biosystemseng.2022.12.001
13. Dias PA, Tabb A, Medeiros H (2018) Apple flower detection using deep convolutional networks. Comput Ind 99:17–28. https://doi.org/10.1016/j.compind.2018.03.010
14. Alyafei MAS, Al Dakheel A, Almoosa M, Ahmed ZFR (2022) Innovative and effective spray method for artificial pollination of date palm using drone. HortScience 57(10):1298–1305. https://doi.org/10.21273/HORTSCI16739-22
15. Link H (2000) Significance of flower and fruit thinning on fruit quality. Plant Growth Regul 31(1–2):17–26. https://doi.org/10.1023/a:1006334110068
16. Vanbrabant Y, Delalieux S, Tits L, Pauly K, Vandermaesen J, Somers B (2020) Pear flower cluster quantification using RGB drone imagery. Agronomy 10(3):1–26. https://doi.org/10.3390/agronomy10030407
17. Barnett J et al (2017) Robotic pollination—Targeting kiwifruit flowers for commercial application. Int Tri-Conference Precis Agric, no June 2020
18. Zawbaa HM, Abbass M, Basha SH, Hazman M, Hassenian AE (2014) An automatic flower classification approach using machine learning algorithms. Proc. 2014 Int Conf Adv Comput Commun Informatics. ICACCI 2014:895–901. https://doi.org/10.1109/ICACCI.2014.6968612
19. Ahmad I et al (2022) Deep learning based detector YOLOv5 for identifying insect pests. Appl Sci 12(19). doi: https://doi.org/10.3390/app121910167
20. Pinheiro I, Aguiar A, Figueiredo A, Pinho T, Valente A, Santos F (2023) Nano aerial vehicles for tree pollination. Appl Sci 13(7):1–26. https://doi.org/10.3390/app13074265

21. Tubau Comas A, Valente J, Kooistra L (2019) Automatic apple tree blossom estimation from uav rgb imagery. Int Arch Photogramm Remote Sens Spat Inf Sci—ISPRS Arch 42(2/W13):631–635. doi:https://doi.org/10.5194/isprs-archives-XLII-2-W13-631-2019
22. Oppenheim D, Edan Y, Shani G (2017) Detecting tomato flowers in greenhouses using computer vision. Int J Comput Inf Eng vol 11(1): 104–109 [Online]. https://publications.waset.org/10006411/detecting-tomato-flowers-in-greenhouses-using-computer-vision
23. Xu R, Li C, Paterson AH, Jiang Y, Sun S, Robertson JS (2018) Aerial images and convolutional neural network for cotton bloom detection. Front Plant Sci 8:1–17. https://doi.org/10.3389/fpls.2017.02235
24. Horton R, Cano E, Bulanon D, Fallahi E (2017) Peach flower monitoring using aerial multispectral imaging. J Imaging 3(1). doi:https://doi.org/10.3390/jimaging3010002
25. Hulens D, Van Ranst W, Cao Y, Goedemé T (2022) Autonomous visual navigation for a flower pollination drone Machines 10(5):1–17. https://doi.org/10.3390/machines10050364
26. Dias PA, Tabb A, Medeiros H (2018) Multispecies fruit flower detection using a refined semantic segmentation network. IEEE Robot. Autom Lett 3(4):3003–3010. https://doi.org/10.1109/LRA.2018.2849498
27. Aggelopoulou KD, Wulfsohn D, Fountas S, Gemtos TA, Nanos GD, Blackmore S (2010) Spatial variation in yield and quality in a small apple orchard. Precis Agric 11(5):538–556. https://doi.org/10.1007/s11119-009-9146-9
28. Aggelopoulou AD, Bochtis D, Fountas S, Swain KC, Gemtos TA, Nanos GD (2011) Yield prediction in apple orchards based on image processing. Precis Agric 12(3):448–456. https://doi.org/10.1007/s11119-010-9187-0
29. Hočevar M, Širok B, Godeša T, Stopar M (2014) Flowering estimation in apple orchards by image analysis. Precis Agric 15(4):466–478. https://doi.org/10.1007/s11119-013-9341-6
30. Teodorescu G, Moise V, Cosac AC (2016) Spatial variation in blooming and yield in an Apple Orchard, in Romania. Ann "Valahia" Univ Targoviste - Agric 10(1):1–6, 2016. doi: https://doi.org/10.1515/agr-2016-0001
31. Liu L et al (2023) Influence of different liquid spray pollination parameters on pollen activity of Fruit trees—pear liquid spray pollination as an example. Horticulturae 9(3):350. https://doi.org/10.3390/horticulturae9030350

Printed in the USA
CPSIA information can be obtained
at www.ICGtesting.com
LVHW022023291223
767724LV00005B/458